Critical Visualization

Critical Visualization
Rethinking the Representation of Data

Peter A. Hall
Patricio Dávila

BLOOMSBURY VISUAL ARTS
LONDON • NEW YORK • OXFORD • NEW DELHI • SYDNEY

BLOOMSBURY VISUAL ARTS
Bloomsbury Publishing Plc
50 Bedford Square, London, WC1B 3DP, UK
1385 Broadway, New York, NY 10018, USA
29 Earlsfort Terrace, Dublin 2, Ireland

BLOOMSBURY, BLOOMSBURY VISUAL ARTS and the Diana logo are trademarks of Bloomsbury Publishing Plc

First published in Great Britain 2023

For legal purposes the Acknowledgements on p. 6 constitute an extension of this copyright page.

Cover design Chris Lee
Cover image © 2022

A catalogue record for this book is available from the British Library.

Library of Congress Cataloging-in-Publication Data

Names: Hall, Peter, 1965- author. | Dávila, Patricio, author.

Title: Critical visualization : rethinking therepresentation of data / Peter Hall, Patricio Dávila.

Description: London ; New York : Bloomsbury Visual Arts, 2022. | Includes bibliographical references.

Identifiers:
LCCN 2022006041 (print) |
LCCN 2022006042 (ebook) |
ISBN 9781350077232 (hardback) |
ISBN 9781350077249 (paperback) |
ISBN 9781350077263 (pdf) |
ISBN 9781350077256 (epub)

Subjects: LCSH: Information visualization--History.

Classification:
LCC QA76.9.I52 H35 2022 (print) |
LCC QA76.9.I52 (ebook) |
DDC 001.4/226--dc23/eng/20220429
LC record available at https://lccn.loc.gov/2022006041
LC ebook record available at https://lccn.loc.gov/2022006042

ISBN
HB: 978-1-3500-7723-2
PB: 978-1-3500-7724-9
ePDF: 978-1-3500-7726-3
eBook: 978-1-3500-7725-6

Typeset by Chris Lee
Printed and bound in India

To find out more about our authors and books visit www.bloomsbury.com and sign up for our newsletters.

Table of Contents

Acknowledgements

The idea for this book began to take shape during a conversation at the Urban Ecologies 2013 conference in Toronto, where we recognised that our research interests intersected, along with a shared sense of a missing critical discourse. A number of events, discussions, and trips helped the project development, for which we would like to thank some very supportive people at our respective institutions past and present: York University and OCAD University (Patricio); Griffith University and University of the Arts London (Peter). A huge thanks goes to our contributors, mentors, mediators and supporters, including:

Danah Abdulla, Janet Abrams, Kim Albrecht, Heath Bunting, Tim Cadman, Lizzie Coles-Kemp, Tom Corby, Tony Fry, Lorraine Gamman, Donna Haraway, Alfredo Jaar, Laura Kurgan, Simon Maidment, Hem Patel, Margaret Pearce, Philippe Rekacewicz, Damon Rich and Jae Shin, Anna Ridler, RNDR, Bronwen Robertson, Rebecca Ross, Jeremy Till, and Rebecca Wright.

Special thanks go to our indefatigable research assistant, Aaron Tucker, illustrator Marc Ngui, and amazing designer Chris Lee. We would also like to thank our editorial assistant at Bloomsbury, Libby Davies, and the editors at Bloomsbury, Louise Baird-Smith, Rebecca Barden and Lee Ripley, who sowed the seed back in 2014. Finally, Peter would like to thank Lana Lê.

P.H. & P.D.

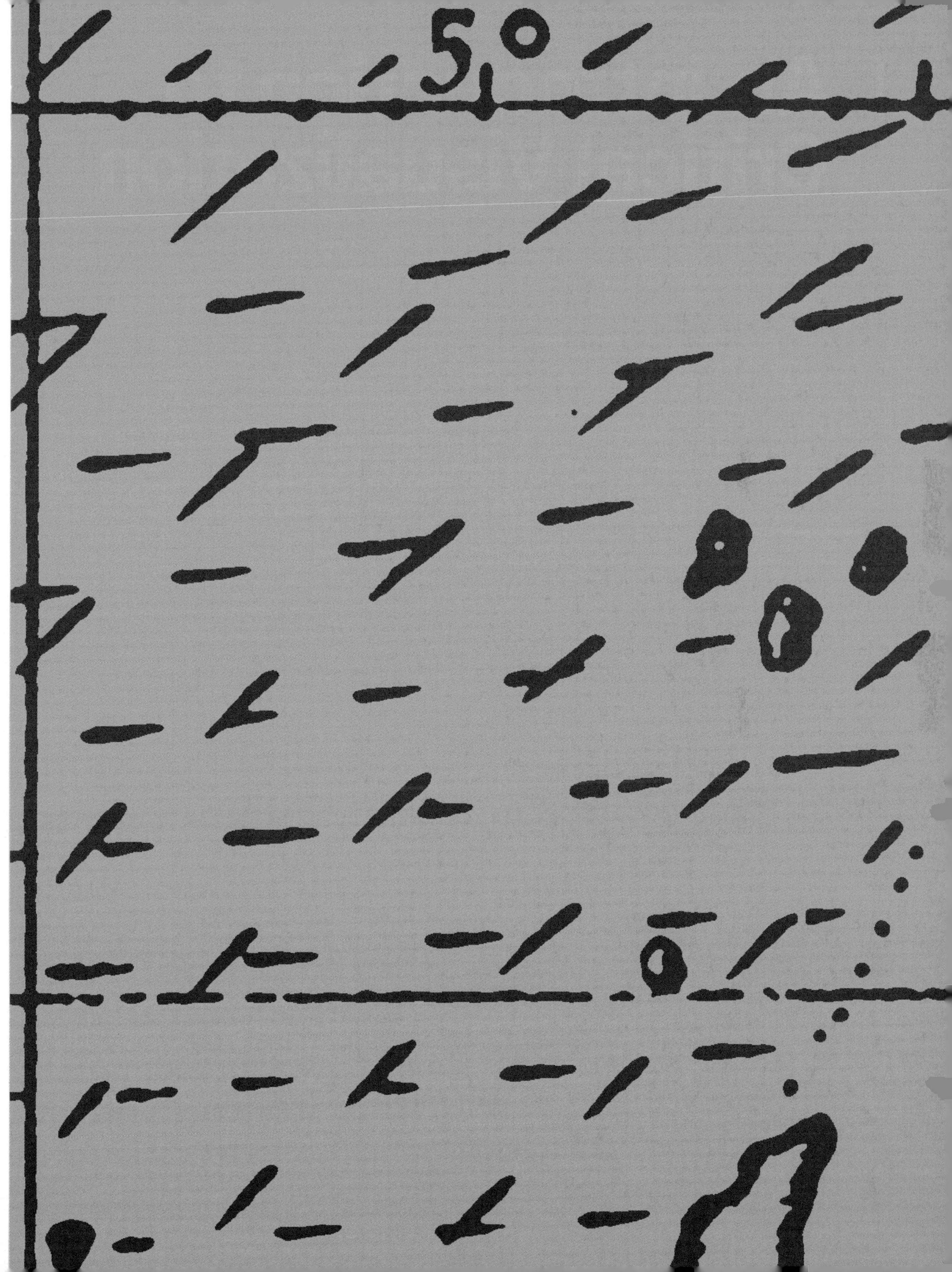
5°

so-called neutral technologies: human decisions sealed behind computational logic (i.e. algorithms). This is one reason why this book makes a case for a critical approach to visualization, to confront a familiar problem exacerbated by the filter bubble effect (Pariser, 2012). On the one hand, there is the blunt deployment of data visualization to support powerful agendas. On the other, there is a blind distrust of visualizations that do not conform to a given prejudice. To counter this effect, we advocate both the rigorous critique and unfolding of data visualization and the practice of creating visualizations that are counter to, or critical of, dominant narratives. To paraphrase the geographer Denis Cosgrove's remarks on maps: always make visualizations, always question visualizations (Ross, 2006).

This book is written for those whose interest, business or practice is the visualization of data, as well as for those who study data and its impact in fields such as human-computer interaction, science and technology studies, critical data studies and digital humanities. As authors, researchers, designers and educators in communication design and media practice, we are interested in how data and its collection, manipulation and presentation both *empowers* and *disempowers* the people and the interests that are represented (or misrepresented) by the data. Our premise is that data visualization has flourished in terms of technique and creativity, but that this has come at the cost of critique. As we shall discuss, the idea of what it means to be critical has itself been critiqued, but we remain convinced that a robust field of data visualization warrants a robust critical discourse.

In the field of communication design and graphic design, the deficit of robust critical discourse becomes particularly apparent when contrasted with related fields such as critical data studies and critical cartography (to which we frequently turn for cues on method). In graphic design circles, as in mainstream media, the burgeoning of data visualization across disciplines has brought with it a form of information fetish or aestheticizing of information that tends to stunt the growth of critical discourse. Entangled with this fetish is the position, influentially demonstrated by Edward Tufte, for example, that the 'efficient', 'clear' and 'excellent' visual display of quantitative information will reveal the truth and help us solve all manner of social and technical problems in a fair, unbiased way (Tufte, 2001, pp. 13–15). The achievements of dataviz proselyte Hans Rosling, for example, included showing how data visualization can counter popular misconceptions about global health, but did not emphasize critical thinking around the circumstances of its gathering, selecting and presenting.

Figure 1.02 — **Hans Rosling presents** ***The Joy of Stats.*** © BBC

A CRITICAL FRAMEWORK

To establish a framework for critical visualization, we might begin with a series of statements, which we explore in the subsequent chapters:

1. **Data visualizations are part of larger *data assemblages*** — complex socio-technical systems, composed of many apparatuses and elements that are thoroughly entwined, whose central concern is the production of data (Kitchin, 2014, p. 24). As we discuss in Chapter 3, a good example of a data assemblage is the census, a large amalgam of apparatuses and elements that shape how a census is formulated, administered, processed, communicated and how its findings are employed.
2. **Data assemblages enhance and maintain the exercise of power** within society, particularly where they assemble institutions, regulations, laws, administrative measures, scientific statements and moral propositions (paraphrasing Michel Foucault's notion of the *dispositif*). One way in which a data assemblage maintains power is in the deployment of categories to organize what, or who, gets counted and how. As discussed in Chapter 3, once a person or group of people is classified, for example, as autistic, affluent, bi-racial or healthy, the classification has an effect on that person or group, which philosopher Ian Hacking terms a 'looping effect'. In this way, data classifications 'make up people' (Hacking, 2006).
3. **Data visualization critique reveals the assemblage**, or *explicates* the assumptions, apparatuses and elements that are entwined in the form: critique does not simply pass judgement on a visualization's formal clarity or excellence (although this certainly helps us make better visualizations). Here we lean on Foucault's account of critique: 'A critique does not consist in saying things aren't good the way they are. It consists in seeing on what type of assumptions, of familiar notions, of established, unexamined ways of thinking the accepted practices are based' (Foucault, 1981/2002, p. 456). In revealing the assumptions and unexamined ways of thinking, critique also reveals what or whose interests are being excluded.

4. **Data visualizations need to be situated in time and space** to counter the visual rhetoric that connotes omniscience, or the 'god trick' (Haraway, 1988, p. 581), that suggests data has been gathered, and is being viewed, from nowhere—a position often confused with *objectivity*. One way to situate a visualization is to draw from the expertise of those familiar with the data site and processes of its generation. We can also borrow a framework developed in Actor-Network Theory to analyse how things take shape through identifying, enrolling and mobilizing other interests. For example, in Chapter 2, we explore how Florence Nightingale's famous polar area graphs developed as part of her process of accruing support from politicians, military chiefs and newspapers, making herself indispensable to the network.
5. **Critical visualization practice uses assemblages, explication and an analysis of power** to create counter-narratives or critical visualizations. This is vital because—as is evident from point 2 of the framework—data visualizations and data assemblages function to enact power and, as such, tend to serve the interest of the powerful. As Catherine D'Ignazio has eloquently argued, 'there is profound inequality between those who are benefitting from the storage, collection, and analysis of data and those who are not' (D'Ignazio, 2017, p. 6).

Three of the terms deployed here—assemblage, situated knowledge, Actor-Network Theory (ANT)—indicate the discourses we are adapting to the task of critical visualization.

The concept of an *assemblage* goes back to Gilles Deleuze's and Felix Guattari's (1983) account of the heterogeneous entities—humans, non-humans, organizations, laws, things—that come together to produce power or agency through their relations.

The *situated* data visualization refers to Donna Haraway's (1988) account of the 'situated knowledges' produced by subjects who acknowledge the contingency of their own position in the world, and, in doing so, produce greater objectivity than those who claim to be neutral.

Actor-Network Theory describes (confusingly) not a theory, but a set of 'material semiotic' tools and methods of analysis for explaining how heterogeneous relations between human and non-human actors are continuously forming and reforming to enact power (Law, 2007). The tools of ANT help us explore how a data visualization translates the disorder of the world into patches of orderliness—by imposing categories, defining, counting and sorting entities into those categories, and then deploying

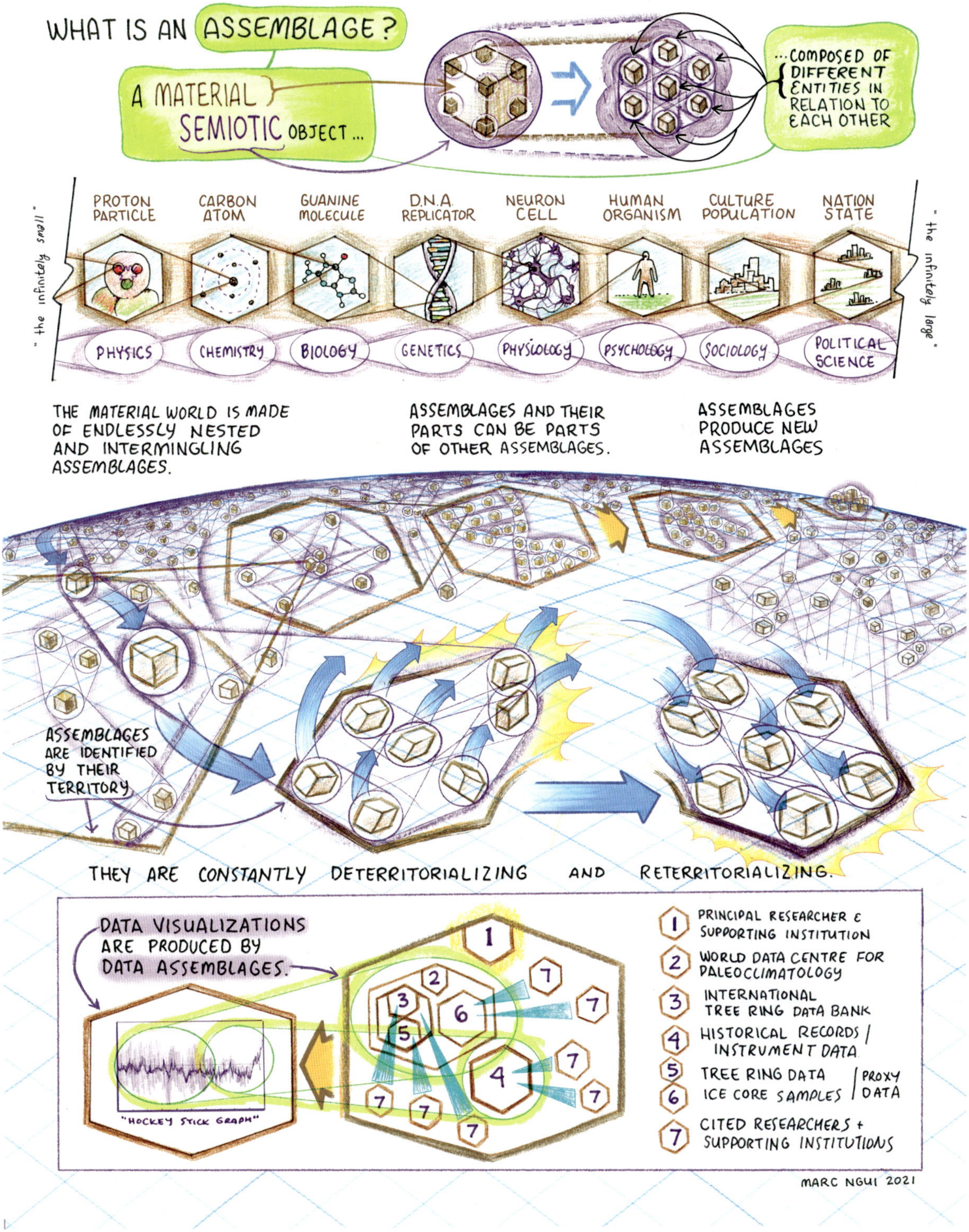

Figure 1.03 — **Assemblage.** Marc Ngui

that order to influence behaviour and shape policy. ANT emerged in the early 1980s and, unlike prevailing literary and linguistic approaches to meaning-making, focused on empirical case studies in science and technology to tease out how social and material things ('actors')—such as electric vehicles, door-closing mechanisms, scallops and Louis Pasteur's laboratory— enacted power by identifying, enrolling and mobilizing social-material relations (Callon, 1986; Latour, 1992; Law, 2007). Because it characterizes power as networks of associations, ANT does not tend to account for factors outside of those associations (e.g. a pervasive ethnocentric bias) that are nonetheless influential (Müller, 2015). Assemblage theory, however, accounts for such a bias, and helps explain how data visualization enacts power.

In short, we are applying to data visualization some of the ideas and methods from the last three decades of scholarship in science and technology studies (STS) and geography (Müller, 2015). Our aim is to show how visualization both *represents* and *enacts* interests, and how it *structures* and *imposes* power relations.

Rather than evaluating a visualization's stand-alone excellence, clarity or integrity, then, we can use the framework detailed earlier to situate and contextualize a visualization's claims to authority and to remind us that every visualization assembles an alliance—or ragtag army—of interests, each with their own history. Krause (2019) sets out a case for writing 'data biographies' with questions probing the authorship, purpose, method of collection, intended audience, impact and limitations of every data set. In a similar vein we can introduce a visualization to our framework with questions loosely derived from the basic investigative method taught to trainee journalists: the *who, what, when, why* and *how*? The sequence that follows leans rather more heavily on the *what*, because, as we shall see, unravelling the conditions of a visualization's development nudges us to consider the ANT-ian 'moments' of identifying, enrolling and mobilizing that accompanied its production and pre-production. So we arrive at *who, when, why, what, what, what* and *what*:

Who made the visualization and for whom?
When did they make it?
Why did they make it?
What social/cultural conditions was it made under?
What belief systems is it reinforcing or challenging?
What processes, or translations, preceded its production?
What has been excluded?

To apply the questions to an example, let's turn to a popular source for the aestheticizing of information as a form of cultural production: David McCandless' much-cited book *Information is Beautiful* (McCandless, 2012) and its cover image, the 'Colours and Culture' visualization.[1]

Who made it? David McCandless, a British data journalist.

When? In 2009, just as 'infosthetics' and dazzling graphics were gaining prominence, fuelled by widely available data mining, scraping and mapping technologies.

Why? The full title, 'Colours and Culture: The Meanings of Colours around the World' indicates that the intention was to demonstrate how colours mean different things in different cultures (e.g. red means good fortune in China, but debt in the West).

In what social/cultural conditions? The reassuring idea that data visualization can help make sense of complex, bewildering, interconnected problems becomes more appealing when an expert guide offers his or her services. As a data journalist, McCandless has developed a practice of data interpretation using spreadsheets, and subsequently design tools, to 'tell stories' with data (McCandless, 2015). Any entrepreneur requires a level of showmanship, and we can speculate that the 'Colours and Culture' visualization was produced more for its form than its function, as we shall explain shortly.

What belief systems is it reinforcing/challenging? On close inspection, the visualization reveals its dependence on tertiary sources, namely Wikipedia and 'general web' (presumably Google). This might indicate the strangely skewed sense of world geography evident in the infographic: there are two continents represented (Africa, Asia), three subcontinents (America, South America and Eastern Europe) two religions (Hindu and Muslim), one Indigenous group (Native American) and two countries (Japan and China).

What has been excluded? There is no Western European ring and the designation 'Asian' is unclear. In fact, the logic behind the categories in the key is unclear, if not entirely ad hoc. In the graphic, 'America' on the outer ring, has 53 entries. 'South America' on the innermost ring, has just five entries (red for danger and success, green for death, purple for mourning, blue for trouble).

1 For a viewable and zoom-able visualization, see p.7 of the book sampler: http://infobeautiful2.s3.amazonaws.com/IIB_Sampler_Dec2016.pdf (accessed 21 May 2021).

The troubling impact of a closer inspection of 'Colours and Culture' is that the visualization's claims to authority, objectivity and even credibility fall apart amid the ruins of its incoherent logic. At risk of lingering too long amid the rubble, we might, prior to departing, consider the apparatuses on which it depended for its appeal to authority: Wikipedia, specifically the world's largest online site of encyclopaedic knowledge production and—presuming that 'general web' means Google—the world's largest search engine. The national, corporate and gender biases inherent in Wikipedia's declared 'neutral point of view' (NPOV) approach to entry-writing have been the subject of extensive debate (e.g. Menking and Rosenberg, 2020). Similarly, the notion that search engines like Google provide a level playing field for 'all forms of ideas, identities and activities' is confronted by Safiya Noble, for example, who exposes the social problem of data discrimination and the oppressive effects of search algorithms (Noble, 2018). One fierce criticism of McCandless' *Information is Beautiful* platform is levelled by designer Neville Brody, who, in a debate on the BBC, argued that 'the idea of making beautiful information misses the point: trying to make information pretty can hide the core message. Information is political' (BBC, 2010). Brody's critique supports our premise that the collection, sorting, arrangement and presentation of data empowers some and disempowers others (point 2 of our framework)—to create a beautiful visualization that purports to show how colours mean different things to different cultures, but to populate it with data gleaned from one dominant region using apparatuses whose effect is to perpetuate a particular epistemology or particular worldview, is to effectively undermine the claim to pluralism.

In the McCandless example, the who-when-why-what questions served to poke behind the glossy facade of a visualization to reveal the larger *assemblage*, or *actor-network*, to which it relates. Law (2007) has argued that there is little difference between the assemblage and actor-network, but while the actor-network approach insists on *symmetry* in exploring the strategic, relational and productive character of heterogeneous entities, assemblage points to the *imbalance* of power that results.

The imbalance of power also provides a rationale and motive for making new, critical visualizations (framework point 5). Those equipped to deal in data generally have power and resources on their side. D'Ignazio identifies the imbalance between the data-haves and data-have-nots in her case for engendering a 'creative data literacy' that benefits the traditionally disempowered.

> Because it is state and corporate actors who possess the resources to collect, store and analyze data, individuals (e.g., citizens, community members, professionals) are more likely to be the subjects of data than to use data for civic purposes. There is a strong case to be made for cultivating data literacy for people in non-technical fields as one way of bridging this gap. (D'Ignazio, 2017, p. 6)

As we demonstrate in Chapter 3, one rhetorical function of a data visualization is to erase or obscure the decisions and machinations that accompanied its development. For the statistical display of quantitative information to be clear and 'excellent', it must eliminate all uncertainty, ambiguity and preparatory work including consultation and participation. Tufte's oeuvre demonstrates this effect, and the process of concealment in visualizing data is strikingly analogous to the way human input is often concealed in algorithmic decision-making. Cathy O'Neil's aptly named book *Weapons of Math Destruction* provides a series of case studies where data and algorithms were deployed, often blithely, to discriminate against their subjects, such as a Washington school district fifth-grade teacher who was fired because her performance was red-flagged by a new algorithm that failed to take into account key contextual information; the teacher went on to a successful career in a private school (O'Neil, 2017, p. 7). Our argument is not to make all data visualization messy and inconclusive, but to draw attention to the rhetorical effect of concealing decisions behind decisive graphical user interfaces.

This book, then, makes a case both for *critiquing* visualizations and for *making* critical visualizations. Both activities draw from the framework above, and in some cases subvert the rhetorical appeals to authority implicit in rules and guidelines on user-friendly or efficient visualization. In both aspects of critical visualization, we take cues from critical cartography, which, as we explore in Chapter 2, emerged in the post-Second-World-War years, out of a discipline that had followed a similar discursive framework as data visualization: a 'cartography of progress' (J. Crampton, 1994, p. 2) governed by internal technical interests and disengaged with ideological concerns (Kitchin and Dodge, 2007).

DEFINING THE FIELD

The field of data visualization is itself subject to different disciplinary perspectives and conceptions of its scope. Disciplines of relevance here are media studies, statistics, computer science, geographic information science and design — all of which have differences as well as commonalities in their approaches to visualization and its critical discourse. In his recent cultural history of data graphics in news media, Murray Dick uses the term 'infographic' interchangeably with 'data visualization', arguing (after Alberto Cairo) that the two concepts exist on a spectrum from

The mechanistic flavour of data visualization discourse is located as a primarily Western bias by Drucker, whose book *Graphesis* makes a well-substantiated case for more interpretive, humanistic approaches to visual forms of knowledge production. Drucker notes that visual forms of knowledge production (i.e. 'graphesis') have played a subordinate role to logocentric and numero-centric forms of knowledge in the West, and a methodological foundation for graphesis has to be 'cobbled together' from an array of contributing intellectual traditions, including: knowledge as vision; languages of form; universal principles of design; Gestalt principles of perception; the semiotics of graphics; techniques of framing and reading; issues of computation vision and the typology of graphic forms. (Drucker, 2014, p. 20).

LOOKING AT VISUALIZATION BEYOND WESTERN PARADIGMS

In his various writings on decoloniality, Walter Mignolo argues for 'delinking' from dominant Western paradigms of thought and practice, and 're-existing' through praxis, the idea that theory is doing and doing is thinking (Mignolo and Walsh, 2018, p. 7).

> Decolonial thinking and doing aim to delink from the epistemic assumptions common to all the areas of knowledge established in the Western world since the European Renaissance and through the European Enlightenment. Re-existence follows up on delinking: re-existence means the sustained effort to reorient our human communal praxis of living. (Mignolo and Walsh, 2018, p. 106)

To delink data visualization from its dominant Western paradigms requires a critical interrogation of the inherent assumptions: whose data? What interests are being furthered in this view, and what interests are being silenced? Here, critical visualization can learn from the work of geographer J.B. Harley, who introduced the concept of 'cartographic silences' — the knowledges buried by the territorial imperatives of the map-makers (John Brian Harley, 2002). Developing new approaches to visualization requires that silenced forms of knowledge are given an opportunity to 're-exist' in new data, in new ways of seeing old data and in new representations of that data.

An example of how a twofold approach to critically thinking about and critically making visualization works is provided by two contrasting projects that responded to the urgent problem of waste generation (according to the World Bank, in 2050, waste generation will reach 3.40 billion tons annually, outpacing population growth by more than double).

The Trash|Track project, launched in 2009 as part of Massachusetts Institute of Technology's (MIT) Senseable City Lab, posed the question 'Why do we know so much about the supply chain and so little about the "removal chain"?' Researchers invited 500 participants to attach an electronic tag to an item of trash, then tracked the movement of these 3,000 objects through the United States, using GPS and mobile phone data. These movements were then represented as elegant visualizations resembling maps of flight paths across the USA, with items colour-coded according to materials (electronics, glass, paper, textiles, etc).

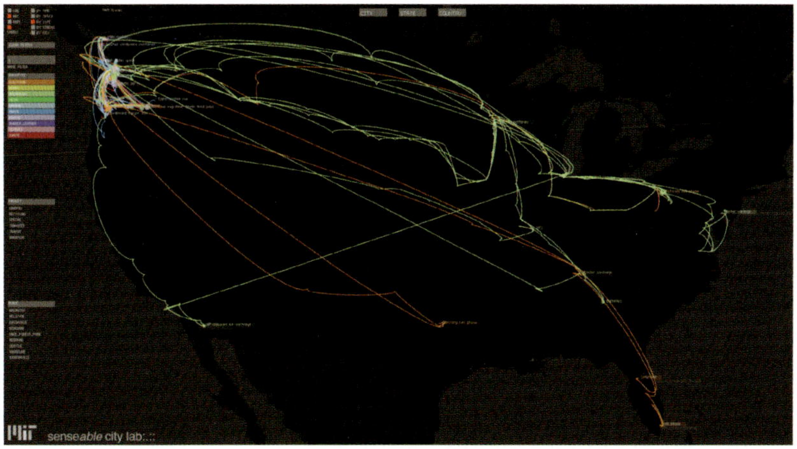

Figure 1.08 — MIT's **Trash | Track project Composite Map.** MIT Senseable City Lab

Figure 1.09 — MIT's **Trash | Track plastic container.** MIT Senseable City Lab

Chapter 4, 'Data and the Self', considers biopolitics and how the self has been constructed by practices of counting, the creation of norms, pre-industrial notions of time discipline and extractivism, and industrial notions of energy and fatigue. We trace the link between measuring bodies and managing bodies, between governance, scientific management, and eugenics in the early twentieth century, and present-day concerns with surveillance and risk profiling. We also consider resistant practices of self-tracking, concluding with a case for satirical and playful practices in the arts, and Henri Lefebvre's (1977/1991) case for the everyday and everyday life as the site of life affirmation.

Chapter 5, 'Data and the City', focuses on smart cities as the locus of a dialectic between the rhetoric of the spectacular visionary 'self-tuning' city and messy, agonistic, participatory forms of mapping and governance. Both kinds of cities deploy visualization, but with dramatically different intentions: the smart city deploys visualization as a tool of persuasion, evoking its connotations of techno-bureaucratic solutions to old and new urban problems (from planning, to traffic congestion, to extreme weather events); the messy, agonistic, participatory and 'accidentally smart' city (Dourish, 2016, p. 37) deploys visualization as a tool for engaging the city's different publics. We consider precedents of the smart city in twentieth-century imaginaries of the frictionless, machine-age city, and look at how, with the advent of climate change, city residents are thrown together in unexpected alliances to work out new ways of being and strategizing everyday life, adapting visualization tools and tactics in a make-do approach.

Chapter 6, 'Beyond Aesthetics and Representations', tackles the patriarchal dismissal of beauty as ornamentation, and its reinforcement by modern minimalist aesthetics and a long-reified signal/noise communication model. We explore the possibilities of beauty as a means of engagement, perception as a means of augmenting cognition, and representation as the re-presenting of selected aspects of lived reality to create meaning—to either uphold or challenge dominant orders. Examples of visualizations that re-present to challenge dominant orders include efforts by W.E.B. Du Bois to create new narratives of African Americans in the early twentieth century, efforts by Bureau d'études to challenge totalizing views, and efforts by Iconoclasistas to subvert the stereotypes inherent in the long-revered Isotype visualizations by Otto and Marie Neurath. To harness its disruptive, revolutionary potential, visualization needs to be considered as an action rather than an object, and the agency of designers needs to be recognized as a force to shape the world they aim to represent.

Chapter 7, 'Beyond Critical Visualization Practice', concludes the book with a provocation for delinking data visualization from Western paradigms, commencing with a critique of the same critical theory that renders universal a Western understanding of the world. Here, we support Glissant's (1997) case for opacity over transparency, since the former acknowledges the irreducible complexity of things being visualized and refuses to rest in the conceit that transparency (the logic of letting the data 'speak for themselves') and simplicity (think of graphic design's beloved 'big idea' approach) should be upheld as paradigms. In place of the pursuit of universalist and rationalist visualizations produced through top-down, synoptic views, we support local, situated, Indigenous perspectives, oral traditions, knowledge of place and *vincularidad*. In place of the saviour complex and pain-centred research (e.g. 'I have visualized and mapped your problems, and you need this'), we support visualization produced with and for the people whose interests it represents, following Eve Tuck (2009): desire-centred research, and visualizations that act.

DISTRIBUTED COGNITION AND HUMANISTIC APPROACHES

We have positioned this book as an introduction or provocation, because it targets an area we feel is in need of development, while acknowledging that as researchers embedded in the field of design—in UK and North American universities—we are, ourselves, learning what we have to learn. This includes the rich discourse that challenges the dualism at the heart of the fields of human-computer interaction and visualization. Better known as Cartesian dualism, named after the seventeenth-century French philosopher René Descartes' famous meditations on first philosophy (1647), which set out reasons to 'doubt about all things and especially about material things', the dualist/Western tradition prioritizes theory over action, and mind over body. Descartes established a foundational philosophical method, but stopped short of using it to question his most fundamental assumption, that his language privileged a particular worldview. By imagining that an 'evil genius' had conjured up a world that appears real, for the purpose of setting traps for his credulity, Descartes then established proof of an 'I' with the affirmation that there must be an 'I' who is capable of doing all of the doubting: *cogito ergo sum* (I think therefore I am). Non-Western traditions, however, do not cleave to the conception of an 'I' that precedes action, that is autonomous and somehow apart from the world.

The connection between the Cartesian tradition and visualization is apparent in the way that maps are conceived as artefacts to represent 'things, concepts, conditions, processes, or events in the human world' (Harley and Woodward, 1987), and diagrams are conceived as plans for action in the world, or as descriptions of states encountered in the world. Because the implied order of events is linear—we create a diagram, map, or visualization usually *before* we proceed to act—a priority is assumed (i.e. theory before practice), along with a non-reciprocal model of cognition: the mental image, or visualization, is formed, and *then* our experience or data is organized into its structures. Mignolo and Walsh's impulse to 'disobey' this conceit is supported by anthropological and psychological perspectives on navigation.

Anthropologist Tim Ingold draws from psychologist James Gibson's 'ecological approach' to visual perception to argue that humans do not construct representations of the world inside their heads and then proceed to act on them. Instead, humans use a tacit and sensorial practice of 'knowing as you go' (Ingold, 2000). At the time Gibson published his work on perception (1950–1979), psychology assumed that people's cognitive systems organized the raw material of experience into an internal model that would, in turn, provide a guide for action (Ingold, 2000). Gibson took issue with the implicit separation of mind from the environment:

> Perception, Gibson argued, is not the achievement of a mind in a body, but of the organism as a whole in its environment, and is tantamount to the organism's own exploratory movement throughout the world. If mind is anywhere, then, it is not 'inside the head' rather than 'out there' in the world. To the contrary, it is immanent in the network of sensory pathways that are set up by virtue of the perceiver's immersion in his or her environment. (Ingold, 2000, p. 2)

If the human mind is reconceived as a network of sensory pathways, rather than a central processing unit—to adopt the computational analogy—the implications for data visualization are quite profound. Conventional approaches to data visualization (e.g. Card et al., 1999; Ware, 2004) tend to see the tools of visualization (e.g. diagrams, working out maths problems on a notepad) as 'amplifying' cognition, with the underlying assumption that cognition takes place 'in the head':

> This approach—often called 'reductionist'—largely ignores the environment, assuming that after we have a better understanding of the individual cognizer, we will be able to understand how cognition interacts with the complexity of the environment (Liu et al., 2008, p. 1173).

In advocating an alternative 'humanistic' approach to visual knowledge production, or graphesis, Drucker (2014) characterizes the deficiencies of the reductionist approach to graphical interface design and the 'science' of human-computer interaction (HCI):

> The HCI 'user' combines two ideological illusions in a single paradoxical identity: the predictability of a mechanized automaton and the myth of autonomous agency. Humanistic approaches to interface need to recuperate the theoretical formulation of subjectivity as a part of the enunciative apparatus, of positions spoken, articulated, created by the structuring and desiring machines of representations. The legacy of a half century or more of theoretical discourse is available for this work, ready to be brought back into play. Who is the subject of an interface? How are we produced as subjects of the discourses on the screen? And in our embodied and culturally situated relations to screens and displays? (Drucker, 2014, pp. 146–147)

Drucker proposes that we can also think of interface as an 'ecology' or a 'border zone between cultural systems and human subjects' (Drucker, 2014, p. 148). A humanistic design approach, she argues, is 'subject-oriented' rather than 'user centred'. Drucker contrasts statistical graphics with humanistic graphics, whose forms and lines express uncertainty, and situated data that supports more complex processing models to gain meaning:

> Interpretation is stochastic and probabilistic, not mechanistic, and its uncertainties require the same mathematical and computational models as other complex systems. (Drucker, 2014, p. 131)

Rather than expect typologies and formats of graphic elements to account for the ways in which meaning is produced, Drucker builds a case for an interpretive understanding of how meaning is produced by readers or viewers, citing Erving Goffman's frame analysis as a particularly relevant model for understanding how, in contemporary digital media, we are constantly shifting frames to understand what types of information are being offered, what tasks are made possible and which behaviours are expected.

An alternative approach to HCI's dependence on the autonomous and anonymous 'user' as a unit of study was introduced by the anthropologist Edwin Hutchins with distributed cognition. In his 1995 study of US Navy personnel aboard large ships, Hutchins showed that cognition is a distributed system, both internal and external to the cognizer. Recounting the emergency procedures enacted aboard a ship losing steam drum pressure in San Diego harbour, Hutchins observes that 'many kinds of thinking were required' to bring the ship to safe harbour: 'some quite clearly both inside and outside the heads of the participants' (Hutchins, 1995, pp. 5–6).

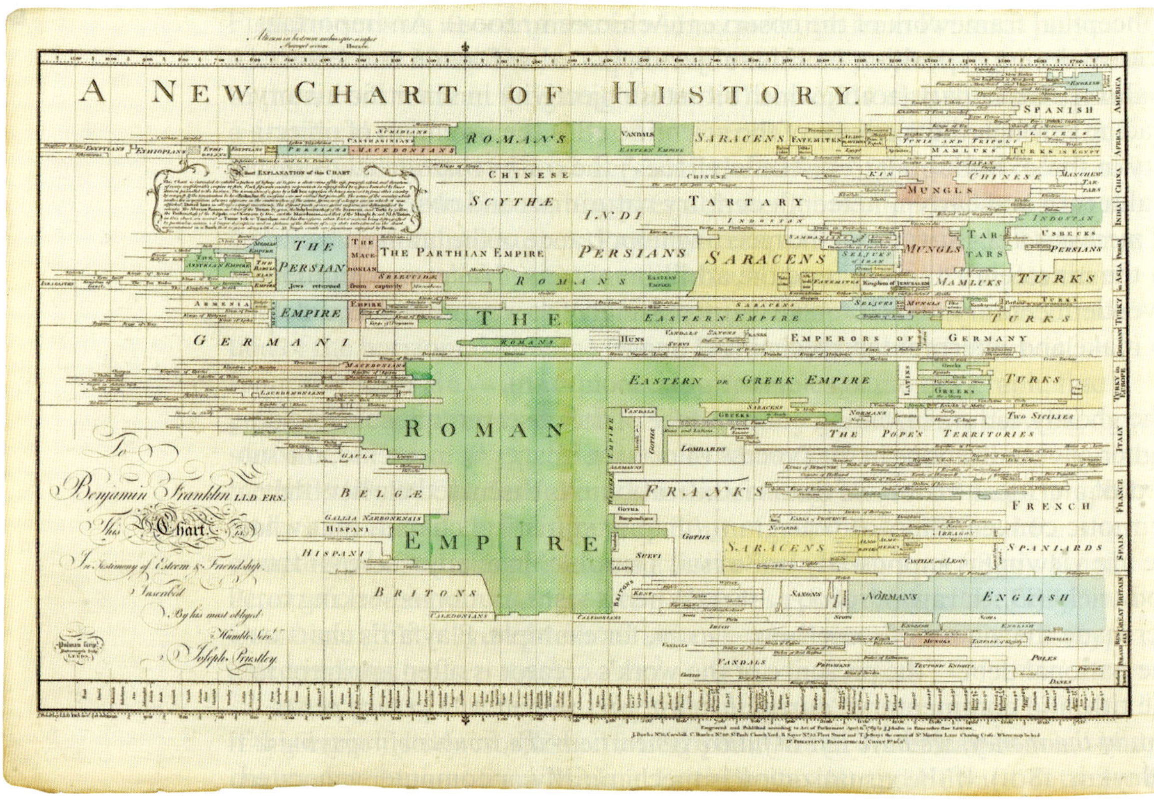

Figure 2.01 — **Joseph Priestley's linear timeline of history (1769).** Joseph Priestley

A HISTORY OF PROGRESS

The second characteristic of existing data visualization histories is their linearity, their adherence to a narrative of scientific progress. This is a position ultimately inherited from the anthropocentrism of Enlightenment thinkers like John Locke, whose philosophy invoked a 'golden age' when humans began the exploitation of nature's unending bounty through the development of technology (Markley and Cohen, 2012, p. 55). In contemporary accounts, data, like nature, is often portrayed as a resource that can be 'mined': visualizations are successful when they communicate 'what the data has to say' with efficiency, that is, they give the viewer 'the greatest number of ideas in the shortest time with the least ink in the smallest space' (Tufte, 2001, p. 51). The efficient presentation of data then becomes a yardstick with which to distinguish the milestones worthy of canonization.

In Friendly's 'brief history' in the *Handbook of Data Visualization*, significant 'milestones' of data visualization meet criteria under one of the following classifications:

→ contributions to the development and use of graphic forms
→ graphic content
→ technology and enablement
→ theory and practice
→ theory and data on perception of visual displays
→ implementation and dissemination
(Friendly, 2008, p. 8).

Conventional histories of data visualization, like any history of a technology, are *teleological* in the sense that they are unidirectional and assume an endpoint. This position is sometimes implicit in visualizations themselves, for example in Priestley's *Chart of Biography* (1765) and *New Chart of History* (1769) (in D. Rosenberg and Grafton, 2010, pp. 18 and 115), which continued to be printed into the nineteenth century. So ingrained is the metaphor of the timeline today that it is quite difficult to imagine time flowing in other directions, despite the contention that any interesting narrative moves both backwards and forwards in time and digresses into subplots and contrasts. The philosopher Henri Bergson, writing at the time of Marey's chronophotographic gun, considered the metaphor of the timeline to be a 'deceiving idol', and advocated the concept of 'duration' (time as people subjectively experience it) in opposition to the dominance of measured (clock) time (D. Rosenberg and Grafton, 2010, p. 23). Nevertheless, Priestley was convinced by the evidence of his own selection of figures and events in his two charts to argue that the mass of entries on the right of the chart represented the 'acceleration' of the arts and sciences in his time (T. E. Rosenberg, 2007, p. 68).

The teleological flavour of the history of data visualization is evident in its stone-skipping account of innovations, which typically occur with more frequency as we near the present. History with an emphasis on the survival of the fittest techniques is to be found, for example, in Arthur Robinson's (1982) *Early Thematic Mapping in the History of Cartography*, which tracks the flurry of carto-statistical activity in the nineteenth century with the availability of statistics, and the introduction of colour printing and lithography. Michael Friendly similarly describes a 'golden age' of statistical graphics and thematic cartography in the late 1800s, only surpassed by innovations of the more recent post-war 'rebirth'. Certainly, evidence is bountiful of a mid-nineteenth-century European and North American interest in visually presenting numbers and distribution of populations, occupations, religions, the movement of goods and people, and 'moral statistics'—along with a series of technical innovations for representing the data: bar charts, pie charts, histograms, line graphs, time series plots, contour plots and scatter plots (Friendly, 2008, p. 4). Dot maps, for example, were introduced to represent totals with the size or number of dots, choropleth maps rendered given districts in flat tones, patterns or colours to comparable numbers (density, ratio, departure from average) and isoline maps represented common quantities with connecting lines (Robinson, 1982, p. 111). But this well-established progressivist narrative has yet to be dismantled or challenged by non-Eurocentric or non-teleological histories.

To disrupt a seemingly innate sense of historical teleology requires that we look more closely at the assumption, adapted from Darwin's survival of the fittest, that the techniques and technologies that are successful are—by definition—the best.[2] To critique this teleological position, the philosopher Michel Serres uses the example of the modern-day car, which is not the end result of a series of technical improvements on previous technical improvements, but a 'disparate aggregate of scientific and technical solutions dating from different periods'—a Neolithic wheel, 200-year-old thermodynamics and a global positioning system from the 1990s. Serres describes linear history as 'narcissistic':

> We conceive of time as an irreversible line, whether interrupted or continuous, of acquisitions and inventions. We go from generalizations to discoveries, leaving behind us a trail of errors finally corrected—like a cloud of ink from a squid ... I cannot help thinking that this idea is the equivalent of those ancient diagrams we laugh at today, which place the Earth at the center of everything, or our galaxy at the middle of the universe, to satisfy our narcissism. (Serres and Latour, 1995/2008, p. 48)

CRITICAL CARTOGRAPHY: A 'DEFINING MOMENT'

Diagrams which place the present day at the end of a line of human technological progress, evoke a sense of self-importance similar to that of world maps that place Europe at the centre, following the Mercator projection of 1569. In this respect, the critical cartography initiatives of the mid-twentieth century present an invaluable example of how a visualization can critique a hegemonic one, and can disrupt a dominant, unexamined way of thinking. The Mercator projection, like any attempt to render a sphere on a two-dimensional surface, is guilty of distortion: by removing the poles and lowering the equator, it gives more prominence to the northern hemisphere. By placing Europe at the centre, the projection literally placed at the centre of the world the seats of the empires that pro-

2 '... as though today's world was the precise goal toward which all decisions made since the beginning of history, were consciously directed' (Ferguson 1974b, 19). For critique of this assumption, and a case for non-linear accounts of technological developments, see for example Pinch, T and Bijker, W. (1987) 'The Social Construction of Facts and Artifacts: Or How the Sociology of Science and Technology Might Benefit Each Other'. In *The Social Construction of Technological Systems*. Edited by Wiebe Bijker, Trevor Pinch and Thomas Hughes. MIT Press.

ceeded to colonize the rest. Critical cartography's 'defining moment' (to use geographer Jeremy Crampton's phrase) followed the publication in 1974 of Arno Peters' equal-area projection, which, unlike the Mercator projection, enabled accurate comparison of continent sizes (J. Crampton, 1994). Although the Gall-Peters Projection (as it became known) would seem to belong in a history of thematic mapping, Robinson did not include it in his account (Robinson's own projection had been endorsed by the National Geographic Society as the closest of all flat maps to render the round surface of the earth 'without distortion' (J. Crampton, 1994)). In fact, Robinson launched a sustained attack on the Peters projection, taking issue with its claims to universality, while conspicuously ignoring its declared ideological aim of challenging the bias in the Mercator projection (J. Crampton, 1994). Robinson's critique of the Peters Projection curiously included an appeal to taste, arguing that the map was 'reminiscent of wet ragged long winter underwear, hung out to dry on the arctic circle' (1982, p. 104). As discussed in Chapter 6, aesthetics can guide our reading of a visualization, and references to taste, standards of beauty or elegance can influence our perception of value.

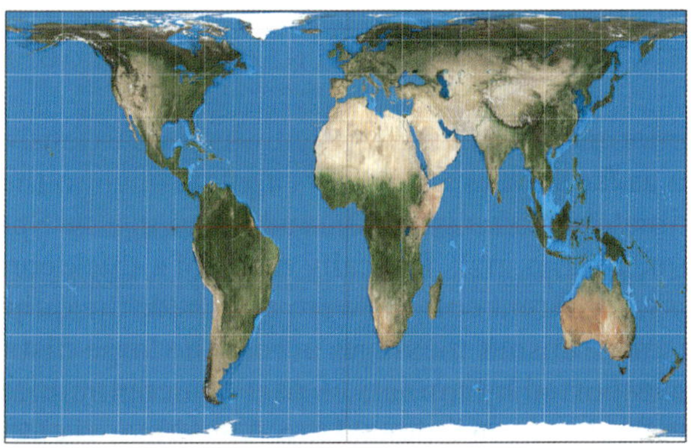

Figure 2.02 — **Cartography's 'defining moment': The world map using the Peters Projection.** Daniel R. Strebe, licensed with CC BY-SA 3.0.

Prior to the Peters controversy, the discipline of cartography, like that of data visualization, was 'overwhelmingly governed by internal technical interests, and did not engage external ideological interests ... it was in fact a "cartography of progress"' (J. Crampton, 1994, p. 17). Post-Peters controversy, key journals such as *Cartographica*, books and conferences paid increasing attention to the relationship between cartography and society. Among the influential work was that of the geographer J.B. Harley, who began applying post-structuralist techniques to 're-read' the text of a map, deploying Foucauldian and Derridean approaches to deconstructing the map and the 'progressive paradigms' of cartographic history (J. B. Harley, 1988).

as an object both foundational to the nursing profession and to a statistical project that privileged colonial power. This privileging is evident in our next case study.

GREAT TRIGONOMETRICAL SURVEY OF INDIA (1802–1875)

The imposition of an architectural masterplan on populations in order to impart a ruling hierarchical order and 'civic values' is a practice that extends back through the history of city planning and mapping, but with the British colonization of India, a particular approach to information architecture was deployed to support the subjugation and categorization of local populations. The Great Trigonometrical Survey (or GTS, 1800–1871), undertaken by the British East India Company in the nineteenth century, imposed a conceptual image of British rule in the visualization and implementation of 'imperial space' (Edney, 1993, p. 61).

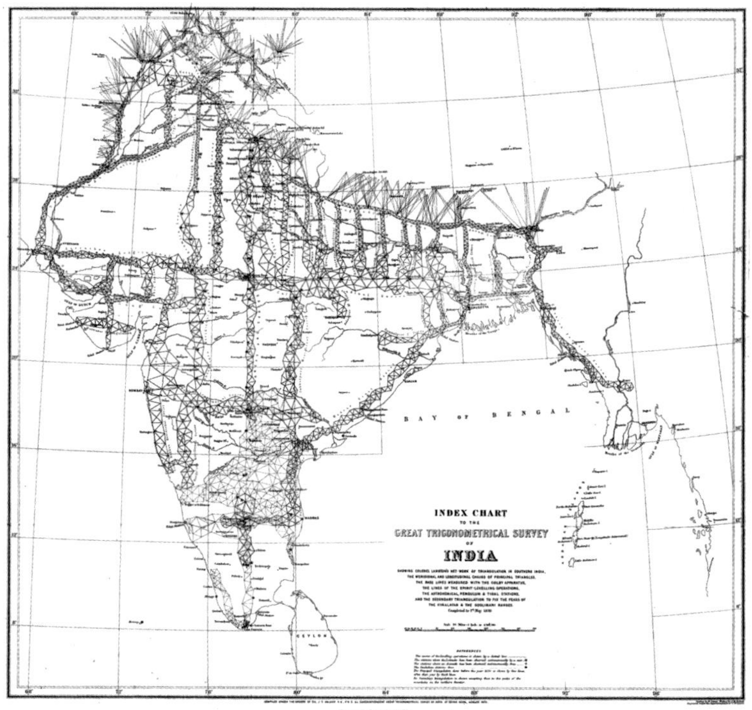

Figure 2.06—**The Great Trigonometrical Survey of India (1802–1875)**. Survey of India

Prior to the establishment of British rule in India after the Indian Rebellion of 1857, British interests in the region were represented by the East India Company, which led a surreptitious colonization of the continent based on trade, brutality and the establishment of bridgeheads in the coastal cities, from which point it administered an increasing number of territories of India and wielded a private army larger than that of the British Army (Dalrymple, 2015; Kulke and Rothermund, 2016). The GTS was a means of mapping the vast and diverse territory of India with a uniform, geodetic triangulation survey, whereby a surveyor establishes a baseline between two known points and determines the location of a third point by measuring the angles to it from the baseline points. By overlaying on the territory a mesh of interlocking large-scale 'primary' triangles, the surveyors could establish a skeletal framework from which to create smaller-scale triangulations. This, in theory, imposed a coherent system for mapping the vast, culturally and politically diverse area as a single entity: subordinates were directed to survey resources (forests, mines, soil), commerce, populations, religions, languages, local histories, animals and medicines (Edney, 1993, p. 62). In reality, the expense and complexity of triangulation meant that the GTS was considerably more ad hoc than the British rhetoric suggested; its completion and supposed coherence became a matter of principle, as a symbolic representation of the supposedly rational and advanced state of European science, compared with the perceived 'indefinite ... imprecise or completely false' knowledge and conceptions of space held by Indians (Edney, 1993, p. 63).

As a visualization, the GTS presents an image of India that appears controllable, reflecting the proliferation of systematic territorial and statistical surveys of the era, which, as Edney notes, homogenized or obliterated local understandings:

> The application of the same techniques and scales of enquiry to each and every district meant that the resultant maps and statistical tables all contained the same sorts of information and were constructed and tabulated in the same manner. They therefore obscured or denied local nuances and particular circumstances. (Edney, 1999, p. 36)

A parallel project was the population census of India, which the British commenced in 1865. While appearing to be a passive instrument of data gathering, the census created, by its practical logic and form, a new sense of category identity, becoming an instrument of knowledge and a form of social control. As Arjun Appadurai writes, 'numbers gradually became more importantly part of the illusion of bureaucratic control and a key to a colonial imaginary in which countable abstractions, of people and resources at every imaginable level and for every conceivable purpose, created the sense of a controllable indigenous reality' (1996a, p. 117). Over time, Appadurai notes, the system was co-opted and turned against the colonial power. Counting Indian bodies, as Appadurai puts it, unyoked social groups from the complex structures and agrarian practices in which they'd been embedded, and ultimately was turned into the idea of self-rule, as nationalism became a mass movement. If you have been counted, you count. As discussed in Chapter 4, counting and categorizing people came to depend on a notion of the 'average' or 'normal' inherited from the desire of Quetelet, among others, to establish universal statistical laws. With the colonization of distant lands, entire populations were perceived as *not* average, i.e. savage and unruly, raising the stakes for the Empire's need to govern by imposing a system.

DATA VISUALIZATION AT THE PARIS EXPOSITION, W.E.B. DU BOIS (1900)

The sixty hand-drawn charts, graphs and maps produced under the direction of African American sociologist and activist W.E.B. Du Bois for the 'American Negro exhibit' in the Palace of Social Economy at the 1900 Exposition Universelle in Paris are unusually prescient, in form and content. Formally, the visualizations suggest a creative re-imagining of the bar charts, pie charts and timelines that had begun to proliferate in Europe and the US after their introduction in the late eighteenth century. Among the works were: colour-coded maps showing the distribution of African American populations and land ownership; tables with images and percentage bars showing the breakdown of income and expenditure among 150 Black families in Atlanta, Georgia; and inventive spiral graphics showing property ownership and rural/urban population distributions of African Americans in Georgia and the US as a whole. Created by a team of students working under Du Bois's direction, the visualizations suggest an awareness of precedents by Playfair, Minard and Nightingale, combined with a reinvention of the forms that creatively anticipated the work of Russian Constructivists nearly two decades later.

The context of this unusual and—until recently—neglected event in the history of critical visualization begs elaboration. Du Bois, a professor of sociology at Atlanta University and the first African American to earn a doctorate, was tasked with putting together an exhibit for the Paris Expo by the African American lawyer and educator Thomas Calloway, who had successfully belatedly raised $15,000 from the United States Congress to fund an exhibit of the 'educational and industrial progress of the negro race in the US'. With just four months to produce the exhibit, Du Bois and Calloway commissioned and curated a set of maps, charts, 500 photographs, reports, and a collection of magazines, books and ephemera to combat a number of prevalent myths about Black people in America.

Such myths included the misappropriation of Darwinism to apply it to race, to argue that Black inferiority would inevitably lead to that population's extinction (Battle-Baptiste and Rusert, 2018, p. 33). Data sourced from the census and Atlanta University sociology lab in fact showed that the Black population of the USA was increasing, and illiteracy was 'less than that of Russia, and only equal to that of Hungary' (Levering Lewis and Willis, 2010, p. 29). Calloway characterized the prevalent prejudice: 'Everyone who knows about public opinion will tell you that the Europeans think of us a mass of rapists, ready to attack every white woman exposed, and a drug in civilized society' (Levering Lewis and Willis, 2010, p. 16). Rather than giving lectures on Black Americans that Europeans wouldn't attend, or publishing in newspapers or books that Europeans wouldn't buy, argued Calloway, there would be no more effective response to such slander than to showcase the best in African-American life at the Exposition Universelle, where thousands of Europeans would be in attendance. He also enlisted the support of Daniel A.P. Murray, the assistant librarian of congress (and only the second Black American to work for the library)[4] to assemble a bibliography of 1400 titles by African-American authors, including Du Bois's *The Suppression of the African Slave Trade to the United States*.

The exhibit set out to subvert conventional perceptions of African Americans by presenting a 'mirror image' (Levering Lewis and Willis, 2010) of the white world with photographs of Black Americans, mostly taken by Thomas Askew, Harry Shepherd, and Frances Benjamin Johnston, of middle-class ministers, teachers, lawyers, skilled labourers and students. But with the visualizations, Du Bois also sought to 'blur the colour line', by showing how perceived differences between races were due to factors 'other than inherent racial deficiencies' (Bini, 2014, p. 58). With the exhibition organized in four parts: history, contemporary social and economic conditions, educational progress and production of literature, Du Bois focused the visualizations on the state of Georgia, which had the largest population of men, women and children of colour (860,000) and showed population sizes, proportions of slaves and freemen prior to 1870, proportion of African American children in school, distribution of Black Georgians in the state, net worth of households, acres of land owned and their value, and taxes paid on property. Such progress was set against the backdrop of socially imposed hardship, represented in the exhibit by a map of the slave trade, timelines including the emergence of the white

4 https://www.loc.gov/resource/rbc0001.2013gen06011/?st=gallery (accessed 1 June 2022).

supremacist Ku Klux Klan and lynchings, and an excerpt from the discriminatory Black Code of Georgia (a symbol of disenfranchisement of African Americans after the Reconstruction period).

In effect, Du Bois subtly used the data visualizations at the Paris Exposition to situate historically the progress of Black Americans, framing the advancement of education, professions and populations in the context of centuries of slavery and oppression. Evaluated through the conventional lens of a statistical history of data visualization, the unconventional forms of the hand-rendered gouache and ink charts and diagrams present a challenge: the choices of shapes seem whimsical and some of the colour schemes counter-intuitive. An assessed valuation of taxable property, for example (Figure 2.07), attempts to visualize an increase in the value of property 'owned by Georgia Negroes' from $5.4m in 1875 to $13.4m in 1900. The chart takes the form of a circle, with the lowest dollar amount assigned to a prominent centre circle painted black and the highest amounts assigned to the thinner surrounding circles. Visual logic suggests the width of each circle represents the proportion of gain on the previous year, but the functional effect of narrower width rings is to make it impossible to clearly render the assigned dollar amounts. Du Bois's team chose to place the valuation figures on arrows protruding from the outer rings that pierce the centre circle. The visual effect suggests an invasion or encroachment of the centre circle. Similar design decisions hamper the readability of one of the spiral diagrams on city and rural populations,[5] which draws attention to a red centre representing 734,952 'Negroes living in the country and villages', but visually deprioritizes the total 123,000 or so who have moved to cities, by representing them as a single-coloured line, moving away from the centre at sharp angles.

5 See Library of Congress Prints and Photographs division, https://www.loc.gov/resource/ppmsca.33873/

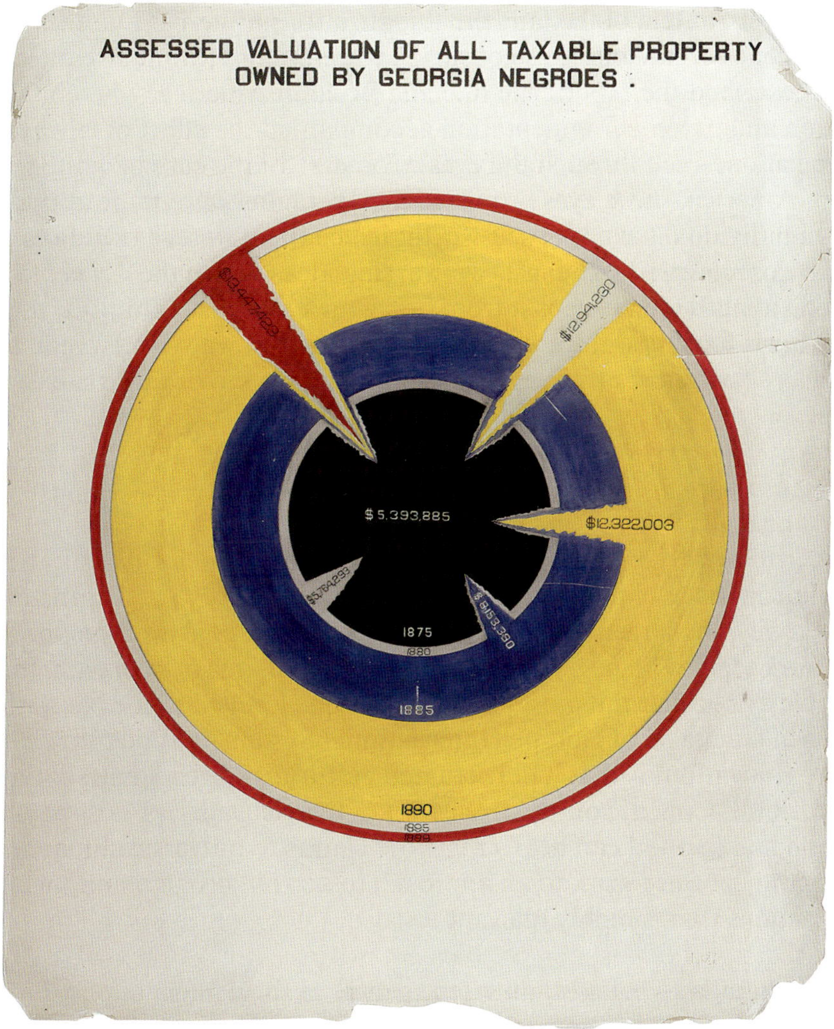

Figure 2.07—**Du Bois's distinctive chart of taxable property owned by African Americans in Georgia (1900).** W. E. B Du Bois

In the broader context of a World's Fair, however, where colonial exploits were traditionally presented as European triumphs, and Indigenous people were othered or exoticized as being 'of nature' with no history of their own, the visualizations suggested an authoritative and rigorous perspective on the living, education and economic conditions of African Americans. The exhibit also presented a unique perspective of the African-American experience, one that must have struck a chord with jurors, who awarded the installation several prizes. In contrast with contemporaneous visualizations used to classify populations, economic trade, poverty, and disease for the purpose of government and management, the hand-rendered forms commissioned by Du Bois created an identity for its subjects. In Chapter 6, we explore how these 'data paintings' re-presented Black Americans.

In the larger context of the fair, the visualizations hinted at Du Bois's famous thesis of 'double consciousness', to be published three years later, which described the experience of being Black in America, whereby social isolation and psychic fragmentation accompanied the effect of 'always looking at one's self through the eyes of others'. In the case of white mainstream America, those eyes projected a disparaging stereotype at odds with an individual's sense of self-worth; hence there were, as Du Bois put it, 'two unreconciled strivings; two warring ideas in one dark body, whose dogged strength alone keeps it from being torn asunder' (Du Bois, 1897). At the Paris Exposition, the struggle against racism was writ large in the setting: a small show of strength emerging out of a sea of white supremacy.

COMMUNITY-BUILDING WITH ISOTYPE: OTTO AND MARIE NEURATH

While Otto and Marie Neurath are commonly credited for developing a formal language of information graphics whose legacy persists in news media today, the larger pedagogical and socially responsive goals of their work tend to be neglected. It is this aspect that is of interest here, particularly because the most common critique levelled at the Neuraths' 'universal language' of Isotype (International System of Typographic Picture Education) is that it perpetuated cultural and gender stereotypes (Bessa, 2008; Cohen, 2014). An account of the Neuraths' philosophy, interests and method can help situate the claims to universalism, and better connect their work to an approach to activist urban planning and interventions that is highly relevant today.

Otto Neurath, a sociologist, initiated Isotype in the context of post-First-World-War Austria, a country so devastated by war its government could not provide the support needed for the nation to recover. Inflation, food and fuel scarcity were rife, and many families set up huts and enclosures in city parks, meadows and forests to grow vegetables and avoid starvation. These 'gypsy settlers' were recognized by many observers—including Neurath—as the seeds of a 'self-help' approach to urban reform and citizen planning. Neurath led a research centre to promote grassroots activism, social economy and self-help housing. The cultivation of 'ordered disorder', as Neurath put it, was necessary, since the actions of citizens driven to fending for themselves were both positive and negative—the emergence of informal economies presented a realistic alternative to market-led capitalism, but needed some coordination and planning. The housing crisis in Vienna, for example, had driven the emergence of squatter settlements, but also threatened anarchy. The shortage of coal had driven families to cut down trees to fuel fires, but this had dangerous consequences for the sustainability of the forests.

The social transformation project underlying Neurath's embrace of pictographic communication is explicated by Nader Vossoughian (2011):

> Neurath was concerned that the industrialized city of the early twentieth century relied on dehumanizing workers, exploiting natural resources, and objectifying human relationships. In its place, he sought to galvanize a global 'polis' premised on the synthesis of community and modernity. (Vossoughian, 2011, p. 12)

Isotype, initially known as the Vienna Method, came out of a belief that the problems of the modern city were social and called for reprogramming and reorganization along the lines of a rationalized in-kind economy that used barter and other 'natural' forms of exchange. After the failure of his initial efforts to support semi-autonomous settlement co-operatives, Neurath turned to museum education and exhibition design as a means to educate and enlist the working classes in the process of transforming the social and economic hierarchies of the modern city. A pre-Isotype example was a series of charts illustrating density levels in cities across Europe and the USA, and the impact of the pace of housing production on the living conditions in a city. A particular target was the 'rental palaces' in Vienna's Ring, which featured ornate and oversized entranceways, but cramped living quarters. 'The study of plans is a beginning,' wrote Neurath. 'They teach us better than any facade whether an apartment is light and airy' (Vossoughian, 2011, p. 67).

Neurath conceived of visual displays that could empower and educate the public on the social and economic processes behind the challenges of the modern city, from housing to production, emigration, unemployment, health and mortality. As founder of the Museum of Society and Economy in Vienna in 1925, Neurath established a 'Department of Transformation', whose role was to analyse statistical data, distil and organize them in an accessible way. The mathematician Marie Reidemeister (who later married Neurath and continued Isotype after his death) took on this job, and the artist Gerd Arntz was tasked with creating the graphic forms. While the linocut pictorial signs of Arntz are well-known today for their influence on information graphics, the role of the *transformer* developed by Marie Neurath is a less-noted but significant aspect of the design process, since it involves the framing, selecting and organizing of data—analogous to James Corner's account of the 'doubly operative' processes of mapping: 'digging, finding and exposing on the one hand, and relating, connecting and structuring on the other' (Corner, 1999, p. 225).

Today, the role of the transformer is often assigned to a data journalist, but as Robin Kinross (2009) observes, the Neuraths' small team established the vital and interdisciplinary aspects of the role through a systematic research process. An early visual diagram of police interventions in Vienna from February 1925, for example, depicts seven rows of unruly-looking individuals divided by a central axis, those on the left representing the number apprehended under the influence of alcohol, those on the right not under the influence. This is a visually engaging advance on a typical bar chart display of data. The orientation around a central axis allows comparisons to be made between the two categories, and the organization of data into rows, with each row representing a day of the week, reveals a pattern more meaningful than, say, organization in orders of magnitude: a quick glance reveals that the longest row—the largest number of police interventions—is for Saturday, the day after payday. Yet, Otto Neurath was later critical of how the visual 'livening up' of the data with illustrations of disorderly people had the effect of inviting speculation on the details of individual cases rather than the central theme of the chart, the statistical relations. He wrote: 'If one knows nothing except "arrests under the influence of alcohol" one must make just one type for it and repeat that, as often as the statistical information demands' (Neurath and Kinross, 2009, p. 79). This methodological emphasis on iterative development in the ordering, filtering and arranging of data—the task of the 'transformer'—had a significant impact on the development of data visualization and graphic communication in the subsequent decades.

Kinross makes a convincing case for the influence of Neurath's approach on the London Underground diagram (first produced in 1931 by the electrical draughtsman Harry Beck) which forms the basis of most mass transit diagrams today: by showing connective relations rather than geographical relations, Beck's diagram—like Isotype—eliminated a large amount of detail, to prioritize what was 'really necessary to know' (Neurath and Kinross, 2009, p. 108). Like Isotype, the London Underground diagram employs a strict set of graphic conventions, including the use of colour, to efficiently convey information and connections. An Isotype chart contemporary to the Underground diagram showing the size of armies and casualties of the Great War (Figure 2.08) similarly experiments with forms to facilitate the communicative function of the diagram. Each figure of a soldier represents a million soldiers, with wounded figures colour-coded red and deaths colour-coded black. Symbols are organized into two groups of parallel rows—one group for each side in the War—which makes it possible to easily compare the scale of the armies and the proportion of casualties and mortalities. Most significantly, quantities are not represented by the size of the graphic figure, as was the case in Michael Mulhall's earlier *Dictionary of Statistics* (1883, cited in Neurath and Kinross, 2009, p. 98), but by repetition of the standard unit. As with the Underground diagram, the unconventional configuration of units is determined by the meaning and communicative function (Neurath and Kinross, 2009).

Figure 2.08—**Casualties of the Great War.**
Gesellschafts- und Wirtschaftsmuseum in Wien, 'The Great War 1914-18', 1933. Otto and Marie Neurath Isotype Collection, University of Reading.

Other significant successors to Neurath's systematic approach cited by Kinross are the 'new typography' expositions of Jan Tschichold (1928, cited in Neurath and Kinross, 2009, p. 115)—who worked briefly with Neurath at the Museum of Society and Economy—and Jock Kinneir and Margaret Calvert's system for the UK road signs developed in the 1960s. Tschichold advocated a strict standardized system with given formats, such as letterheadings on A4 with a second colour, red, to underscore information. On Kinneir and Calvert's highway signs, the system is revealed in a single diagram: A unit is defined as the width of a character stem, which is then used to establish the proportions of the directional lines and the positions of names. This has the effect of prioritizing legibility, rather than, say, the size of sign, since the whole ensemble is 'as big as it needs to be: it has no set dimensions' (Neurath and Kinross, 2009, p. 119).

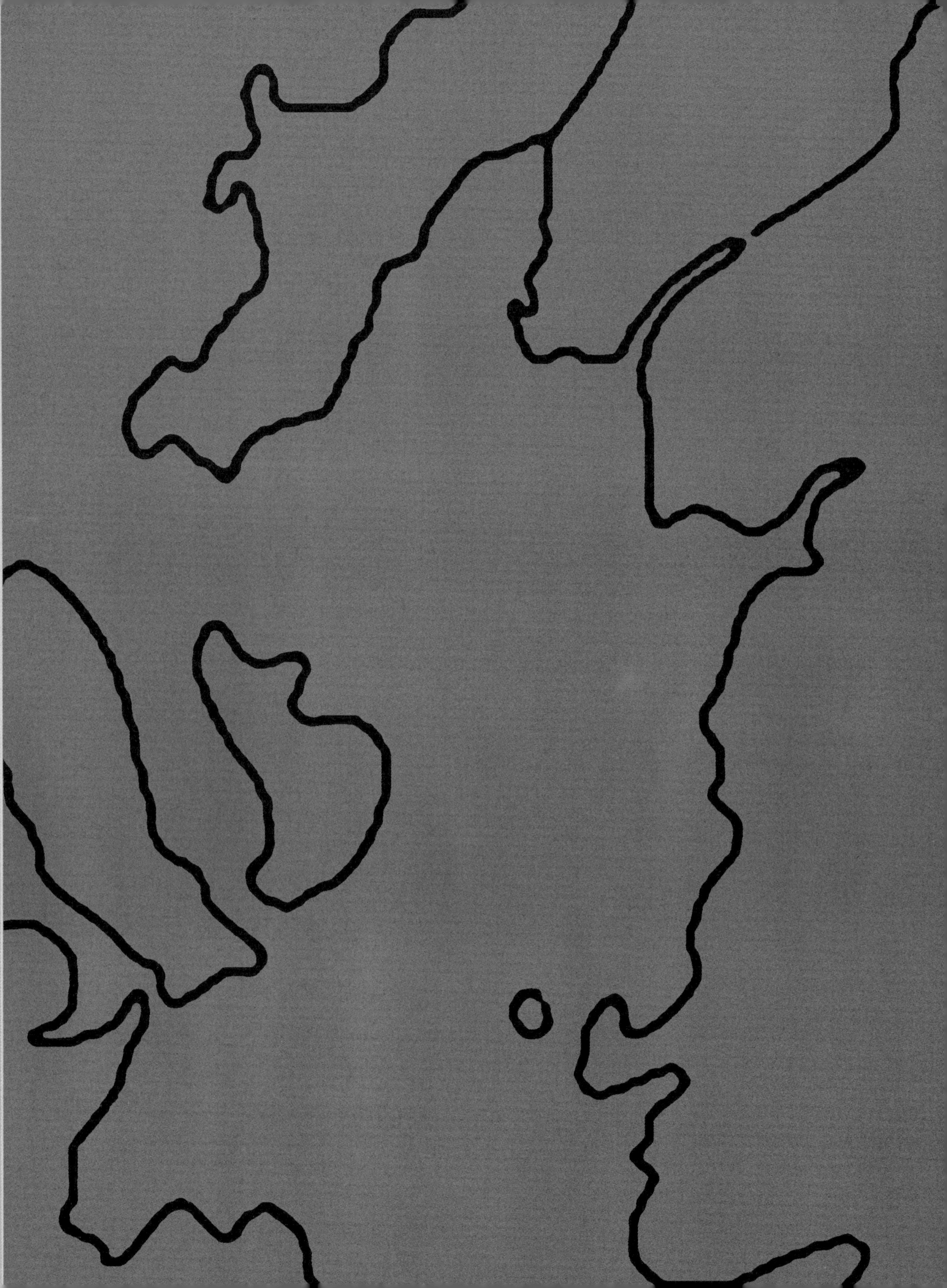

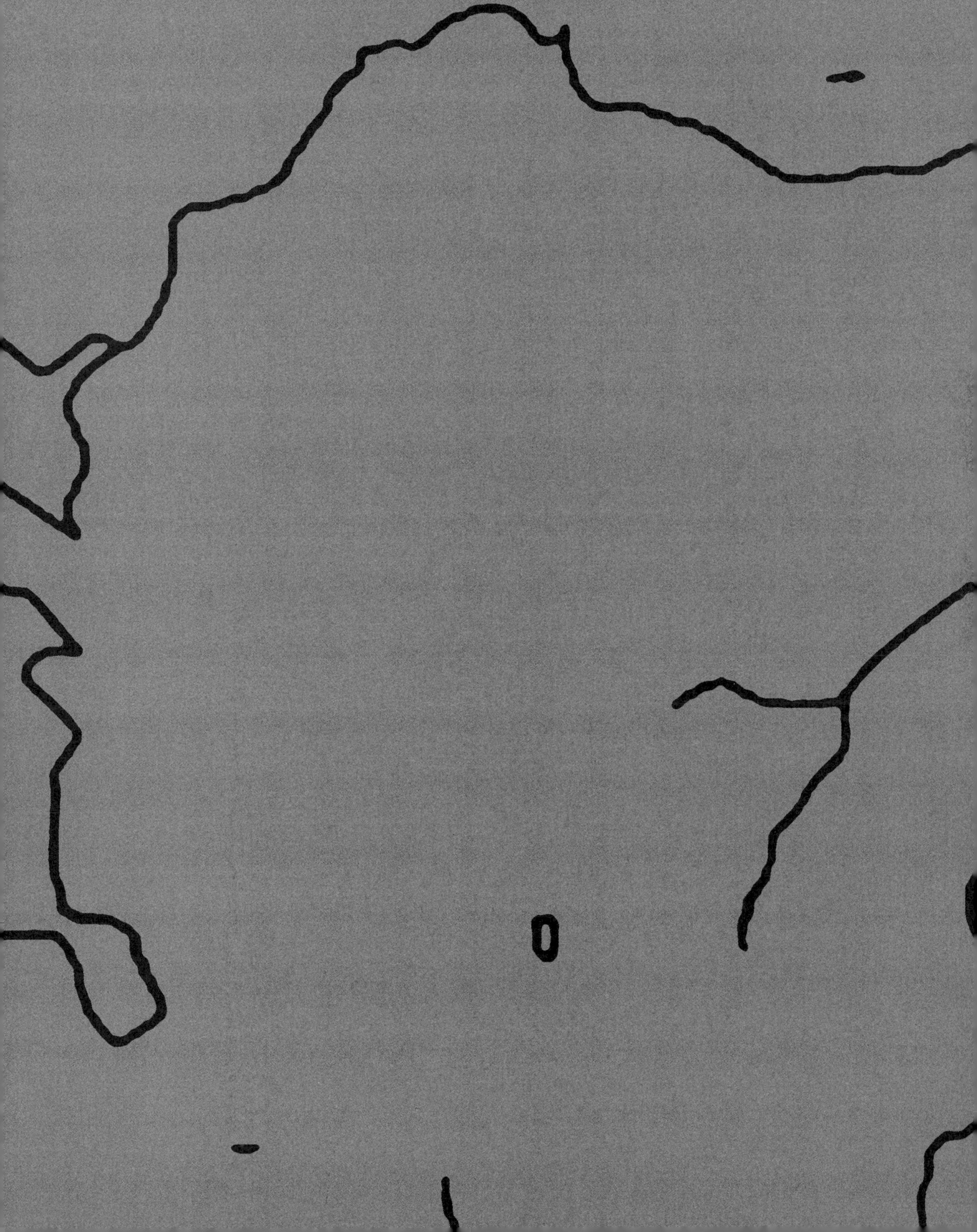

III Making Data

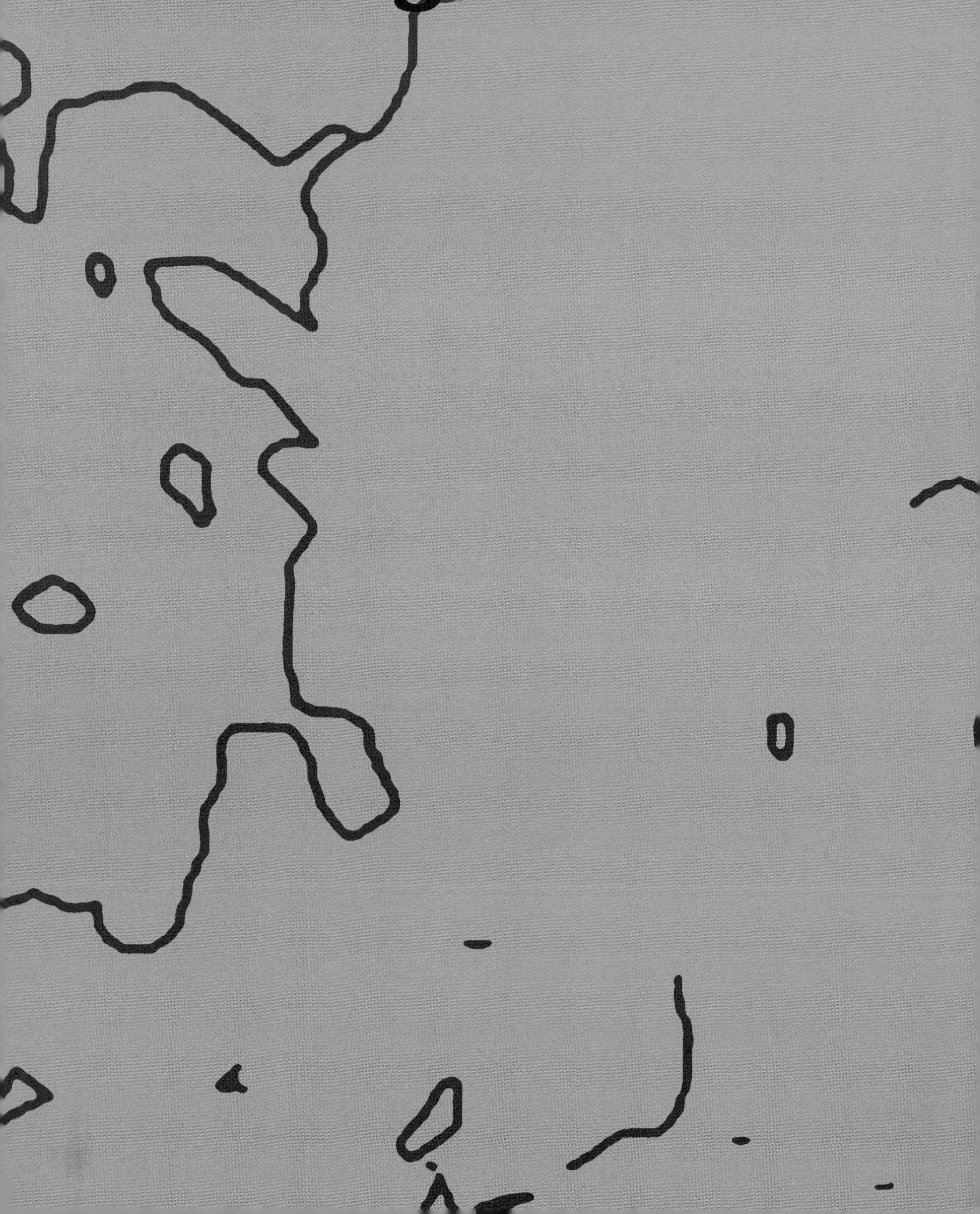

DATA IS A TERM THAT CONJURES IDEAS OF A RESOURCE TO BE exploited or a material to be manipulated. Financial journalists state that data is the new oil (*The Economist*, 2017) and as such, can be extracted and commoditized. The hubris around data is propelled by technological advances in storage and networked devices, as well as the increase in online social life and the corporate strategies to track all transactions. With this enthusiasm for the potential of data to increase profits, there is also an enthusiasm to use data to inform decisions in business, government, research, etc. Visualization, along with other strategies, is used to aid in the analysis, or to represent the analysis, of important data.

As designers working in visualization, we often take data for granted. The visualization of data creates powerful representations of measurements and categories which themselves were created much earlier in the process and often in a context different from that of the designer. While this may mean that how the data is created is outside of the control of the designer, we are still responsible for using this data and reiterating and amplifying its status as a factual representation of reality. For this reason, as critical designers of visualizations, we need to pay attention to the conditions of production of these important components of our projects. Criticality, in Foucault's (1981/2002) account (see Chapter 1), involves paying attention to 'what type of assumptions, of familiar notions, of established, unexamined ways of thinking the accepted practices are based'. As such, criticality can reveal injustice, oppression, misrepresentation and can foreground Haraway's concept of situated knowledge. From this base, this chapter will attempt to address how to think critically about data as it is employed in visualization.

By using data and representing it visually in our projects, we confirm its authenticity unless we explicitly undermine its validity, which is rarely done. We tend to affirm rather than negate the data when we visualize it. This is especially the case when the visualization is clean, minimalist and orderly. By erasing the traces of the parsing, cleaning, filtering and aggregation performed on data we effectively make the viewer forget that the data was *produced*. Or, as Lev Manovich (2002) reminds us, that data is *generated*.

Data visualizations also gain power through how they juxtapose different data sets to draw out a relationship between different phenomena. For data is generated through human-directed processes and is thus subject to being framed in a particular time and place with a particular focus on what is important.

If we look at the etymology of *data* and *fact*, we may see how our current usage entails a particular understanding. The word *data* comes from the Latin *dare* 'to give', and we can contrast this with *fact* from the Latin *facere* 'to make'. While data are seen to be given to us by the world, facts are constructed out of this raw material. Data by definition are 'that which is given prior to argument' (Gitelman and Jackson, 2013, p. 7). We can therefore conceptualize visualizations as attempts to make facts through the use of data. Visualizing thus prioritizes the fact and pushes the data to the background.

The process of visualizing has been modelled as an iterative process (B. Fry, 2008), starting at the acquisition of data, going through a process of analysis, representation and refinement, and ending in a visual and interactive representation. The initial step involving acquisition of data refers primarily to accessing available data, rather than the generation of data. In practical terms, a designer would only need to consider this to begin a project. As this is a model for a technical design practice, the generation of data is not interrogated and is treated solely as a resource to be accessed.

We argue that the production of data is an important aspect of *critical* practice precisely because of the lack of attention normally accorded to it by designers. What is made invisible, in this case, the production of data, is often one of the most important aspects for a critical thinker to focus on. What becomes standardized, routine and efficient has become naturalized, and therefore unquestioned. But it is there, in the unquestioned areas, that so many ideologically, politically, culturally framed decisions have been made. As we stated earlier in the discussion regarding assemblages in Chapter 1, those decisions are made within a power/knowledge discourse that amasses power through the establishment of expertise, circulation of information, and the description of phenomena through standardized measures and categories.

Perhaps it is easiest to imagine this dynamic when we think of the process of measuring and categorizing human populations. Categorization of human beings is very complex and fraught with issues of translating a person's identity or lived experience into a measurable data point. Yet, this same dynamic can be seen within the natural sciences—especially ones that may have controversial implications. The production of scientific data is often challenged on the basis of methodology—that is, on the procedures and protocols used to assure the validity of basic research.

Take, for instance, the measurement of temperature, which in the present era is a very straightforward procedure involving inexpensive, simple and widely available sensors that can provide a quantity in degrees Celsius, degrees Fahrenheit, or kelvin units. If we measure the local temperature over the span of a year, we get a fairly reliable record of how warm it has been by using a stable technology (e.g. thermometer). But if we were to go back beyond one hundred years of temperature data, we would encounter growing inconsistencies of units, unreliable record-keeping and miscalibrated instruments. If we were to go back over a thousand years, the collection of historical data becomes even more difficult. In these cases, instead of modern thermometers, proxy indicators such as tree rings, ice core samples, etc. would be used to record temperature. A visualization showing temperature ranging from present-day to a millennia ago would therefore have to aggregate from several sources and emphasize the continuity of measurement over a long period. Since the focus is temperature, all temperature data are treated as valid sources.

The so-called 'hockey stick graph' is such a visualization that represents the variation in temperatures over a thousand years. Mann et al. (1999) reconstructed global temperatures beyond the era where instruments recorded temperatures and discovered that the last few decades leading up to the year 2000 had been the warmest in millennia. A main innovation of Mann's research and visualization was the bridging of proxy temperature data with instrument temperature data.

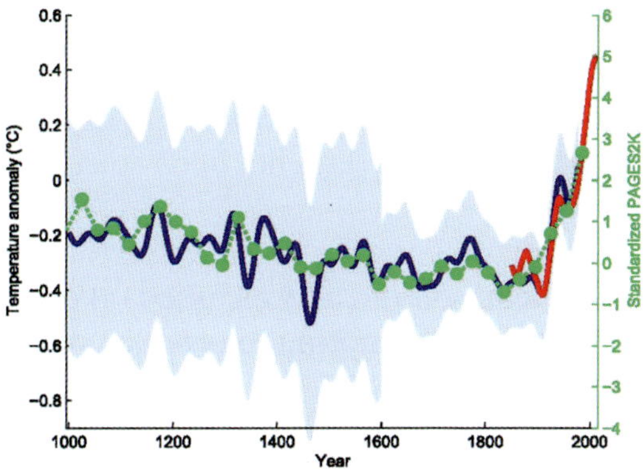

Figure 3.02—**Hockey stick graph (2014)**. Klaus Bittermann, licensed with CC BY-SA 4.0

Essentially, two categories of data were brought together in one visualization. Their graph was assiduously designed to show the different data sources and the margin of error for each data point used. Nonetheless, it became 'a central icon in the climate change debate' (Mann, 2014, p. 22) and as such, a singular image that symbolized one thing: the sharp increase in temperatures over the last one hundred years.

Representing data in visual form is a powerful rhetorical device precisely because it can collapse a complex array of data into a single strong and persuasive image. Visualizations, especially those that give a strong shape to the data, tend to have the effect of erasing the work that led to the making of the visualization. A similar phenomenon has been noted by John Pickles (2004) in the creation of maps, where the pencil marks, mistakes and human judgements are erased in order to present a simplified and objective image. To non-expert viewers, the map or visualization can be extremely convincing because it looks scientific, orderly and objective. In the presentation of observations made through microscopes, we see a comparable process of refinement. Michael Lynch (1991) notes how blurry images that are difficult to distinguish for the untrained eye are progressively simplified in order to display the most salient features. The simplified and annotated images are often the ones that get circulated the most, presumably due to their directness, clarity and contextualizing information. Designing visualizations thus requires an acknowledgement of this powerful feature of data images and a responsibility to ensure the data is valid.

The 'hockey stick' graph, and the analysis it represented, was perceived as a great threat to corporate and political interests which defended the continued use of fossil fuels and denied the human impact on the environment and climate. Many attempts were made to sow seeds of doubt into the analysis of the temperature data and reduce the power of the graph. Climate change sceptics, for instance, split the proxy data and instrument data in order to reiterate that an overall downwards trend in temperature was apparent (Mann, 2014). It is difficult to imagine that this response would have been as severe had the data not been so conclusive and the shape so simple.[6]

The graph that Mann's team produced was based on decades of work done by paleoclimatologists, attempting to produce reliable records of

6 To learn more about the hockey stick graph controversy, read Mann's *The Hockey Stick Graph and the Climate Change Wars: Dispatches from the Front Lines.*

temperatures spanning millennia. Mann describes the production of his, and other climate scientists' work, as being based on an 'extensive global network of long-term proxy data' (Mann, 2014, p. 112). Researchers used tree rings from ancient forests and core samples for the polar regions as the material from which to derive these proxy temperatures. The research has been tested and expanded over a long period of time, and through the work of many international institutions and scientists. This underscores the fact that producing data can also be a very costly exercise, especially when considering scientific data, census data, big data, historical data, freedom of information requests, etc. Latour's (1986) work on the networks of people, processes, and objects that are employed in the generation of scientific knowledge reminds us of the immense amount of work needed to produce data and inscriptions which transmit knowledge, and how this itself makes it difficult to counter these accounts. The criticism that Mann's research received from Western Fuels Association-sponsored climate change sceptics was largely on the basis of technical issues of protocol for entering this research into the IPCC report. The work needed by climate change sceptics to substantively challenge the decades-long research would be difficult to mount. It should be noted that Mann's work, using the same data, has since been independently reproduced several times by different researchers and institutions, which have achieved similar results and made similar conclusions.

Data generation is part of a network of humans, artefacts, and processes that produces knowledge. As discussed in Chapter I, Kitchin and Lauriault call this a *data assemblage*: 'a complex socio-technical system, composed of many apparatuses and elements that are thoroughly entwined, whose central concern is the production of a data. A data assemblage consists of more than the data system/infrastructure itself, such as a big data system, an open data repository or a data archive, to include all of the technological, political, social and economic apparatuses that frame their nature, operation and work' (2018, p. 8). This system is evidenced in the scientific network assembled around the production of climate data in the case of Mann's research. It also helps us understand the parallel network of corporate actors, academics and media organizations that produces counter-arguments, although much less robust or rigorous, which increases uncertainty in the spheres of public opinion and public policymakers. As these interconnected networks demonstrate, facts that are derived from data require work to be maintained and work to retain their validity. This is not to say that they are any less real, but that they are constructed (Latour, 2005) through human (e.g. scientists) and non-human (e.g. trees, sensors) processes.

Figure 3.03—Data generation is part of a network of humans, artefacts and processes which produces knowledge. Here are participants at a Data4Change Design Sprint in Nairobi (2019) tackling digital rights. Data4Change

Foucault's (1980) concept of the *dispositif* is useful here to help understand the stakes involved in this sort of network or assemblage, for it describes the apparatus through which power/knowledge are connected. Power is secured through the establishment of a particular knowledge of a territory, a group of people, a phenomena, an activity, etc. Access to that knowledge and its creation and maintenance is necessarily restricted to those who uphold that power and are invested in maintaining that power. Kitchin and Lauriault (2018) note that data assemblages work similarly. These are infrastructures that operate within a larger context (e.g. political, social, economic) and as such are transformed by them. Shaped by powerful institutions including government, industry groups, academia, scientific or health organizations, these data infrastructures take on many of the agendas (e.g. directions of research) and basic assumptions (e.g. methodology or sources of expert knowledge) that perpetuate their authority over a given field of knowledge.

The difficulty in producing data and its attendant data assemblages also inherently reproduces a divide between those who are part of the assemblage and those who are not. Expert knowledge, access to funding, and technical skills are required in order to produce data. For those without access to such resources, this requirement effectively restricts access to the power of data generation. What emerges is a digital divide in (big) data where powerful organizations produce highly specialized, thoroughly documented, standardized, and widely 'trusted' data sets while non-specialists who are usually at the receiving end of an analysis or policy decision have little access to making data or using it.

Big data exemplifies this dynamic. The complexity and scale of the data requires corresponding infrastructure (e.g. people, hardware, funding) and expertise to manage, use and analyze. One source of big data, social media (e.g. Facebook), amasses this material in order to understand consumer behaviour and advertise to its users. While a user's own data may be downloaded in some cases, the accumulated data of all users is typically available only to the social media corporation and, in an aggregate form as an analytical tool for targeting users for its clients. Another example of this is census data, which is regularly collected by governments every few years in order to collect demographic data within its borders, and therefore, understand issues such as income levels, age distribution, family composition, etc. Traditionally, this data has been collected and analyzed for government planning purposes, with certain derived products or services offered, for a fee, to researchers, organizations and corporations.

QUANTITATIVE AND QUALITATIVE DATA

There are processes that make data conform to a set of conventions in order to allow their manipulation and combination at later stages. Conventionally, in data visualization there are two general kinds of data that are used: quantitative and qualitative. Quantitative data is the kind of data that is in the form of discrete units that can be counted, categorized or analysed via mathematical means. Examples include sensor data, A/B test data, demographic data, etc. Qualitative data is typically descriptive and can be read or analysed via interpretation. Examples include text, sound, image, video, etc. Recorded interviews, journal writing, field notes, photography, or video documentation are the kind of qualitative data that is rich in information. While one kind of data is not inherently better than the other, there are practical and philosophical issues to consider when choosing whether to generate data that can be counted, or data that is descriptive.

In practical application, we need to think of what it is that we are attempting to capture and what is the best way of doing it. A person's height could be described in words (e.g. tiny, giant, tall, regular, big, diminutive, short,

Iconoclasistas and Mark Lombardi perform one important task very often in order to create their maps and visualizations, namely *categorization*. Categorization is a necessary step in the collection and analysis of data. It determines the identity of a particular thing according to a criterion or taxonomy. Bowker and Star define classification as a 'spatio-temporal segmentation of the world' and as 'a set of boxes (metaphorical or literal) into which things can be put to then do some kind of work—bureaucratic or knowledge production' (Bowker and Star, 1999, p. 10).

Yet, difficulties inevitably arise when trying to place things into logical frames such as categories. William Kent (2000) reminds us that entities, which we need to identify in order to place in a category, can be highly ambiguous due to how they are defined by the people who use them. For example, a 'book' may refer to an author's single self-contained work, or a copy of a book or a manual, or a publication of proceedings from a conference. Also, at which point does a book stop being a book? Does the number of pages or the thickness of the cover determine this? The simple term, 'book', reveals itself to be an ambiguous category once we try to establish what is included and what is not.

The process of categorization becomes even more difficult when we try to frame parts of human society into orderly frames. If we take, for instance, the disparity in pay amongst genders, we are required to invoke a set of categories that would underpin such a representation or an analysis. The categories of work, pay and gender are designed through a combination of hegemonic concepts, such as recognized forms of work which often have omitted unpaid domestic work—which nonetheless is a major support for paid work—and concepts such as a male-female gender dichotomy, which erases any representation of trans people.

Nevertheless, categories are extremely useful when identifying things and thinking about the relationships among things. The intuitive understanding of categories is that they contain entities with shared properties. So, for instance, cats are mammals that usually have four legs, a tail, whiskers and chase mice. This also fits the description of dogs. Yet we understand cats and dogs to be separate categories. We can also think of them both as belonging to the pets category. Categories, through human cognitive work, impose frames on the world so that we can think about, and manipulate, the world. These frames are highly influenced by context of use.

Jorge Luis Borges' fictitious taxonomy 'Emporio celestial de conocimientos benévolos' (Celestial Emporium of Benevolent Knowledge) is a good example of how categories impose human values on the world. Written in 1942, it is part of a short piece about John Wilkin, an English man of letters from the seventeenth century, who attempts to devise an analytical language capable of systematically and rationally describing the whole universe. It is noted that the attempt was doomed from the start.

> These ambiguities, redundancies and deficiencies remind us of those which doctor Franz Kuhn attributes to a certain Chinese encyclopaedia entitled "Celestial Emporium of benevolent Knowledge". In its remote pages it is written that the animals are divided into: (a) belonging to the emperor, (b) embalmed, (c) tame, (d) suckling pigs, (e) sirens, (f) fabulous, (g) stray dogs, (h) included in the present classification, (i) frenzied, (j) innumerable, (k) drawn with a very fine camelhair brush, (l) et cetera, (m) having just broken the water pitcher, (n) that from a long way off look like flies. (Borges, 1942/1999)

Foucault (1966/2007), in *The Order of Things*, uses Borges' short story as a reminder of the foolhardiness of the attempt, the arbitrariness of the premise, and the compulsion of humans to categorize things in the world. Lakoff (1990) notes that Borges' passage also reveals a deeper truth about Western attitudes towards non-Western thinking—that it is strange, fantastical and inscrutable, part of an orientalist tradition that treats non-European culture as an *other* that is either romanticized or vilified, or both. This also reproduces a pervasive trope that categories are self-generating, that is, they are evident from the things themselves and it is up to the most advanced scientific expertise (e.g. Western knowledge) to properly name them. Yes, categories are necessary for thinking, but they are relative to their context (Lakoff, 1990) since they are culturally produced and any attempt at producing a universally valid system inevitably erases or marginalizes those things (including people) that do not fit.

Why would categorization be important for a critical visualization designer? If we recall Foucault's concept of the *dispositif* or apparatus, or Kitchin and Lauriault's *data assemblage*, we see the inherent issue of power/knowledge in the creation of data *and in* the establishment of categories. These are processes that structure what we know about the world and its inhabitants and thus have an impact. The human work required to produce data about a human world also means that there is a mutually affecting relationship between that which is classified and that which classifies; Ian Hacking

(2006) calls this *dynamic nominalism*. This is when naming and categorizing result in *making up people*. Identities are literally created to describe parts of society according to particular criteria. For example, people may be categorized as autistic, affluent, bi-racial or healthy. Each of these categories require an individual to possess a set of characteristics in order to qualify. A definitive set of qualities is determined by researchers to be exemplary of that category. The use (in communications, planning, policy, services, etc.) of the language and the category established to describe a person of this kind influences the person described.

Hacking (2006) describes two processes that are at work in this dynamic: *looping effects* and *engines of discoverability*. The looping effect is the process through which data are organized (e.g. classification, the object of focus, the institution, knowledge, various experts) about a given subject (i.e. a person with particular characteristics), who is in turn affected by that description and thus changes in response. On the other hand, the engines of discoverability are the processes that establish a body of knowledge (e.g. biology, medicine, genetics) that attempts to frame the phenomena in its own terms—as Hacking puts it, to *biologize*, to *medicalize*, or to *geneticize*, for example.

Dynamic nominalism is especially salient when we consider how the human world is described. As social-cultural beings, humans are deeply affected by the language and the terms that describe their own existence, which effectively outline the range of possibilities of living their lives. Hacking (2006) uses examples of sexual orientation and psychological conditions which, once categorized and named in discrete terms, inevitably determine a field of experiences heavily shaped by those categories. We can use Hacking's observations to understand how data through categorization and, more specifically for designers, the visual representation of these data and categories, help produce subjects.

As a simple exercise, try to describe yourself by using standardized categories from a census. For instance, Statistics Canada before 2016 used the following single response classifications for identifying race:

1.1 Aboriginal (Inuit, Métis, North American Indian)
1.2 Arab/West Asian (e.g., Armenian, Egyptian, Iranian, Lebanese, Moroccan)
1.3 Black (e.g., African, Haitian, Jamaican, Somali)
1.4 Chinese
1.5 Filipino
1.6 Japanese
1.7 Korean
1.8 Latin American
1.9 South Asian
1.10 South East Asian
1.11 White (Caucasian)
1.12 Other

This would prove difficult if you were Afro-Peruvian, because a respondent can only choose one response—you can select Latin American or Black, but not both. This leads to a significant lack of precision and also an erasure of a particular identity, diaspora or history. Translating from the realm of personal identity and culture that is immensely complex to a domain of countable units leads to the loss of rich information.

If we take from Hacking's observation the effect of categories on people, then we can see how a language (i.e. a system of labels for human identity and experience), which does not include one's own, could affect one's own way of expressing one's identity and experience. These categories are not arbitrarily set, but are produced by a dominant cultural point of view. The official census reflects back to its residents, in this case racial categories, that are based on language already established—not necessarily at the individual or community level but at the hegemonic or dominant cultural level. Furthermore, as Hacking (2006) reminds us, once people are categorized, actions are taken upon them in the form of policymaking, investment, service delivery, etc.

Charles Booth, nineteenth-century industrialist and social policy researcher, developed the Maps Descriptive of London Poverty (part of his *Inquiry into Life and Labour of the People in London*), which detailed socio-economic distribution in the city. Booth's collection comprises twelve to thirteen maps that together form a grid of the city of London in the 1890s.

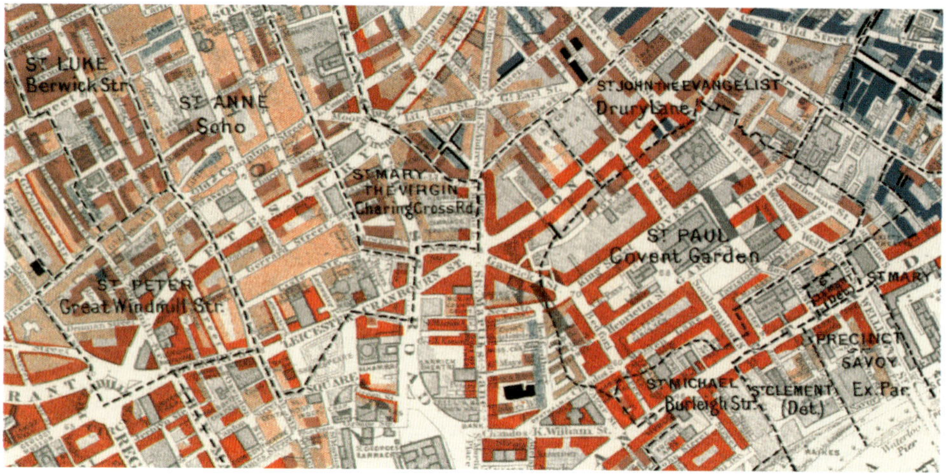

Figure 3.06—**Charles's Booth Poverty Map as digitized by the London School of Economics with original categorization of neighbourhoods.** Charles Booth and 2016 London School of Economics and Political Science

What is significant about this collection of maps is both the scale and the comprehensive nature of the endeavour. For our understanding of data and its function in categorizing human populations it is also significant. Booth used seven categories in his legend:

→ Lowest class. Vicious, semi-criminal.
→ Very poor, casual. Chronic want.
→ Poor. 18s. to 21s. a week for a moderate family.
→ Mixed. Some comfortable others poor.
→ Fairly comfortable. Good ordinary earnings.
→ Middle class. Well-to-do.
→ Upper-middle and upper classes. Wealthy.

Booth utilized the poverty line to delineate a basic division in the city between those who could barely survive and those who could survive at varying levels of comfort. Although extensive research was conducted through surveying neighbourhoods in order to develop a basis for the poverty line (Hacking, 2000), little is known about how exactly Booth established the criterion (i.e. 18 to 21 shillings a week). This categorization did have an effect, as Lauriault (2012) notes, on how poverty was measured in the colonies and the United States—in Canada it directly influenced the use of the Low Income Cut Off used by Statistics Canada.

Compare Booth's work with the identification of Priority Neighbourhoods in Toronto in 2004, later renamed and reconciled as Neighbourhood Improvement Areas (NIA) in 2014. Out of the approximately 140 neighbourhoods in the city, thirteen priority areas were first identified, then this was expanded to thirty-one areas. Initially set up by using socio-economic indicators, proximity to services and the number of annual homicides (Doolittle, 2014), the category was reframed as NIA in order to take into account many more factors, such as health, voter turnout and income level (Doolittle, 2014). The data used in these designations and measurements come from a variety of sources, including the 2011 Census, which collected information on housing, family structure, spoken languages and the 2011 National Household survey, which collected immigration history, ethnic origin, visible minority identification, education and income. These areas are designated *at-risk* and *vulnerable* due to lower income levels, higher homicide rate and poor access to services.

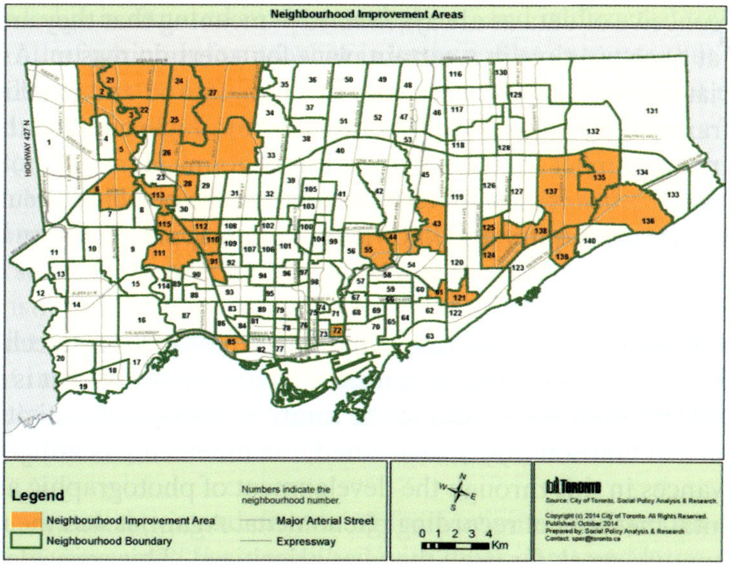

Figure 3.07—**Map of Neighbourhood Improvement Areas.** City of Toronto, Social Development Finance and Administration, Social Policy Analysis and Research, October 2014

Similar to Booth's maps, the data from surveys are mapped onto the physical representation of the city in order to identify concentrations of poverty and thus help direct the government's social and economic policy to improve the conditions and prospects of these residents. Data is generated, reaggregated and analysed for the purposes of governmentality—in other words, to make people governable (Porter, 1995).

Pearce (2014) notes that translation and representation can result in a separation of knowledge from context, which results in a significant loss of meaning. Data, in the form of location and place names, needed to be translated (from its traditional usage and its oral form) in a highly nuanced way, in order to both keep the integrity of the knowledge represented on the map and simultaneously to be accessible to readers who may or may not speak the original Indigenous language. The mapping of Indigenous knowledge highlights the issue of translating complex information (i.e. embodied knowledge that is interwoven with culture and territory) from one domain into another that reduces and separates data from its context.

Criticality in visualization requires that we interrogate the data and information that we are re-presenting in visual form. This interrogation requires that we ask:

→ How was the data produced?
→ How/where is the data production situated?
→ What biases may have framed it?
→ How is the data aggregated and decontextualized?
→ What is gained/lost when it is mapped into other contexts?
→ What set of skills or knowledge is needed to understand the data from its original moment of creation/collection?
→ Who and what is erased when the data is produced?

The definition and context of data can be explored in many ways, each focusing on a discipline, or technology, or usage or policy. In this chapter, we discussed the ways in which data is framed as an assemblage or infrastructure and we also discussed some critical visualization practices that address the ethical ways that data and visualizations are produced. Data does not simply arrive, nor is it something we discover. Instead, we *make* data from people and things that we want to represent. It is a material born out of these relations with sources.

Figure 3.09—**The SizeChina project was initiated to address the exclusion of Asian head shapes in the dominant data set derived from US Air Force personnel.** Roger Ball

The ease with which we can access data and combine data can simultaneously make the system that supports this access and interoperability invisible. When things become invisible they get normalized and less scrutinized for what they reproduce (e.g. what is prioritized and what is omitted). This is precisely the moment when criticality is necessary. The critical stance prompts the questions: 'Who is not here?', 'What has been excluded?', 'Why was it not included?' and 'How is the status quo reproduced?

Data generation is part of knowledge production. It is the generation of material that can help frame a debate or dispel a myth. As designers of visualizations, we choose what data and which stories are amplified through our work. As critical visualization designers, we assume a responsibility for producing those stories and for the ways they were produced. ■

Data4Change

Rather than focus on data visualization *per se*, the London-based non-profit Data4Change works to 'harness the power' of data, in collaboration with local civil society organizations around the world in gathering, organizing and presenting data to achieve lasting change. The group's co-founder Bronwen Robertson started Data4Change in 2014 with journalist Stina Bäcker, based on the knowledge that 'troves of data are locked away in PDF reports that could help create powerful counternarratives to data sets that pretend to speak on behalf of populations' (B. Robertson, personal communication, October 10, 2020). The group has worked on nearly fifty projects in over ninety countries since 2014, including the three projects shown here, which set out to:

1. **visualize how internet shutdowns are impacting human rights,**
2. **advocate for people with visual impairments in Ethiopia, and**
3. **better represent and improve the living conditions of people in Yemen.**

The primary tool for Data4Change is the three-to-five-day sprint, following the UK Design Council's 'double diamond' framework for innovation—discover-define-develop-deliver. However, as the organization's network has grown, it has begun to explore the possibilities of a decentralized network of hubs to increase and amplify existing capacities in the various regions in which it works.

Focus Figure 2.1—Data4Change sprints make sure that data is readable and available for targeted groups and communities. Data4Change

1. KeepItOn

Strategically shutting down internet access has become a political weapon used increasingly by governments in recent years. Access Now, an advocacy organization focused on defending and extending the digital rights of users at risk, established the KeepItOn interactive platform to draw attention to the impact of shutdowns on human rights, businesses, emergency services and journalism around the world. At a Data4Change sprint in Nairobi, Kenya in 2019, the #KeepItOn coalition worked with a creative team to build a new interactive visualization of data from three years of internet shutdown instances and crowdsourced stories submitted to #KeepItOn.

The project has focused on helping Access Now's #KeepItOn coalition develop a sustainable platform for data gathering, by linking shared spreadsheets to the website so that live data can update the visualizations with minimal effort and cost. According to a spokesperson for Access Now, 'the new data entry system has streamlined our data collection and documentation process and, as a result, has increased our data accuracy, reliability and verification process' (B. Robertson, personal communication, October 10, 2020).

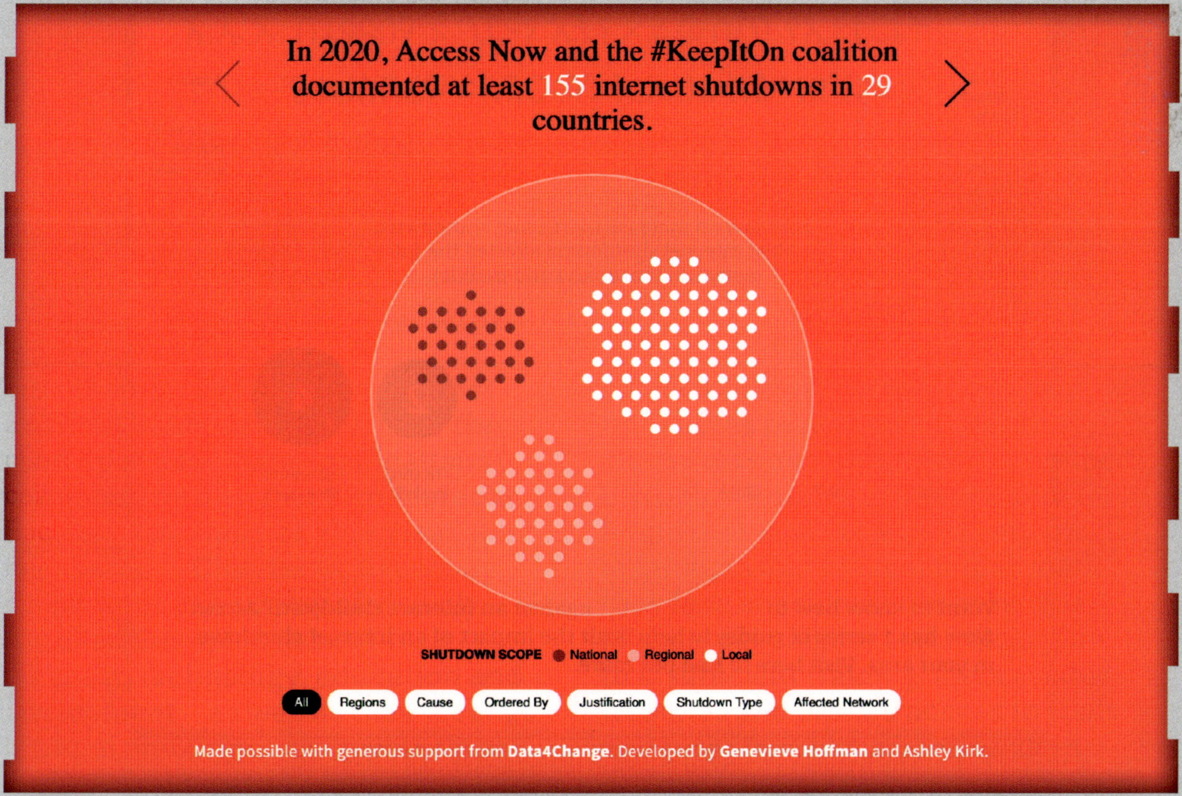

Focus Figure 2.2—Internet shutdowns by region, KeepItOn (2019). Access Now/Data4Change

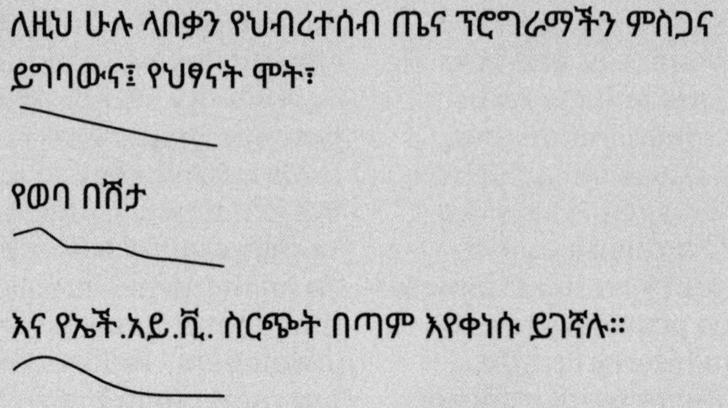

Focus Figure 2.4—Data sonifications were developed to support a campaign about visual impairment in Ethiopia and the impact of digital exclusion. Together! Ethiopia/Kim Albrecht/Data4Change

The sonifications were developed throughout a five-day sprint in Nairobi, Kenya by German designer and researcher Kim Albrecht, as part of the experimental push of a Data4Change sprint. While the various dimensions of visual mapping are at least well defined since Bertin's *Semiology of Graphics* (1967/2011), the translation between data and sound was, at least for Albrecht, still an untold territory. He experimented with mapping numeric values to different audio elements in the same way that data can be mapped to various visual elements. Instead of using colour intensity, size or shape to convey data variables, he experimented with the tone of various instruments, the length of notes, pitch variability, volume levels and trigger attacks. As Albrecht notes, 'The aesthetic of sound is entirely different from images. The mediation of an image functions in space while sound functions throughout time.' (Backer, 2020).

Robertson notes that while the feed-back from a live performance of Hear the Blind Spot at the 2019 forum was very positive, one challenge with data sonification is that it is difficult to achieve aesthetically appealing outcomes while retaining accuracy. 'It works well when you use a single note or pitch to convey the up and down movement of a line graph, e.g. translating the data directly onto pitch, but attempts to then layer in other data using chords or other harmonic elements, can make it overly complicated and lose the specificity' (B. Robertson, personal communication, October 10, 2020). Going forward, the project is being rolled out through more live performances in schools, after delays due to the COVID-19 pandemic. The plan was also modified into an audio broadcast of key health information about the virus that was played via a loudspeaker attached to a roving four-wheel-drive car. 'One key takeaway,' adds Robertson, 'is not to only think about "new" technologies when embarking on data projects, as sometimes it's more effective to "hack the box" and use existing opportunities' (personal communication, October 10, 2020).

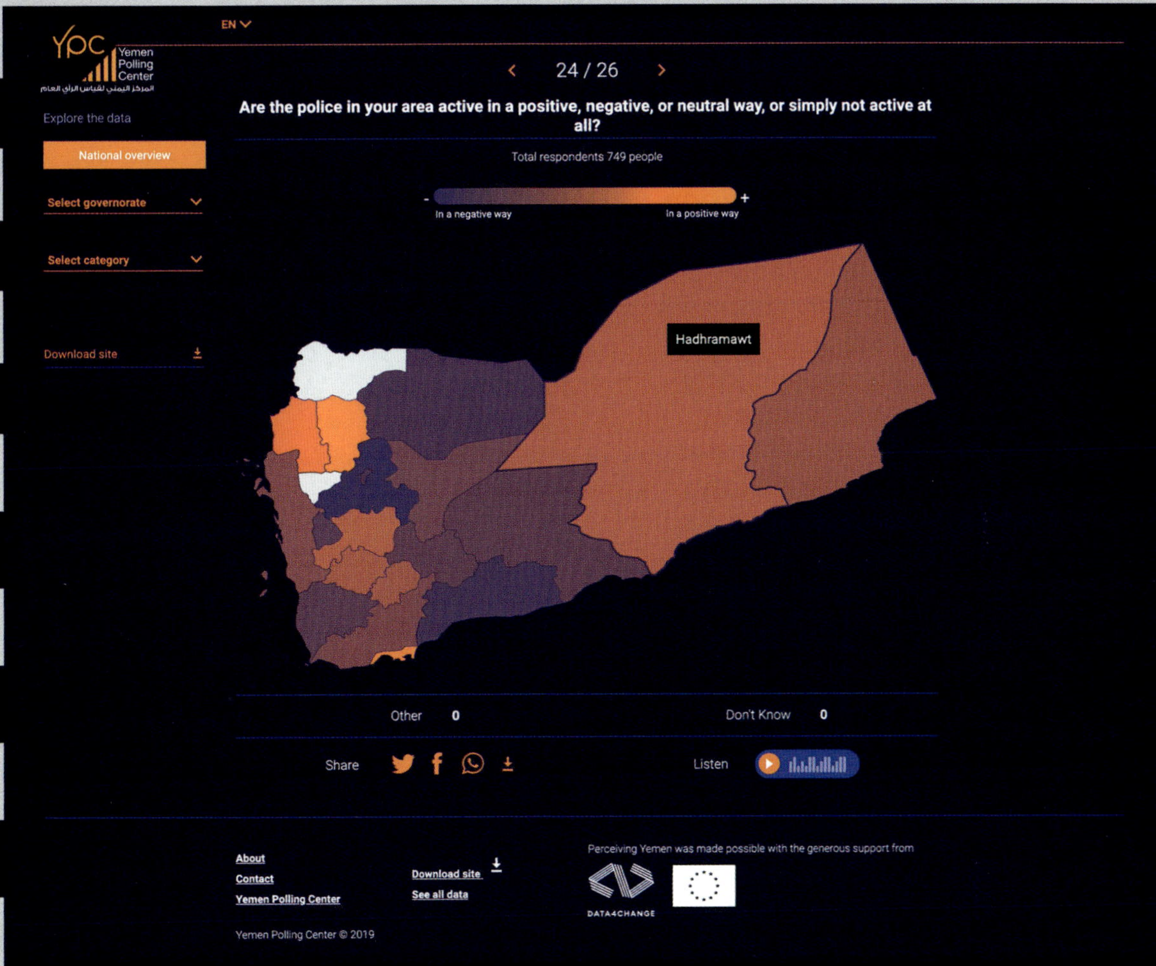

Focus Figure 2.5—Perceiving Yemen is a 5MB downloadable website of data on living conditions across Yemen, designed to minimize energy and data use. Yemen Polling Center/Data4Change

3. Perceiving Yemen

Researchers working for the Yemen Polling Center (YPC), an independent research group established in 2004, face considerable risks in conducting in-person interviews and delivering surveys amid the ongoing conflict. At a 2018 sprint with Data4Change in Bierut, Lebanon, the assembled team explored ways to make two important recent YPC survey results accessible to policymakers and the general public. The first survey, a face-to-face poll of 1,500 young people aged fifteen to twenty-five, included questions on media use and the security situation; the second survey consisted of 4,000 responses from Yemenis about their perception of living conditions and security. At the same event, the sprint team developed a bespoke interface and survey creator, and a system of Wi-Fi hubs as a way to make data collection safer and more efficient (Data4Change, 2020).

At the 2018 Quantified Self Conference in Portland, Oregon, over half of the presentations were on topics related to personal health, diet and exercise: a runner had tracked his sneezes to ascertain which grasses aggravate his allergies; another presenter surveyed the services that analyse DNA swabs to determine health risks. This preoccupation appears to accurately represent the interests of the particular demographic that has taken up the practice of self-tracking, and, perhaps, the economic driver for the movement: the technology. Indeed, the conference exhibit provides an opportunity for toolmakers to showcase their technologies to 'highly engaged advanced users and the self-tracking public' (Quantified Self, 2018).

The question of what is being visualized, and how, begins to reveal motivations. The visual display of personal data ranges from the default outputs of various data visualization softwares, to experimental programmable sculptures, as in Stephen Cartwright's datascapes of his hourly location, in latitude, longitude and elevation. A tendency to present personal data without critical analysis, in the spirit of sharing to gain greater insights, is a Quantified Self-er trait acknowledged and upheld by Wolf: 'For many self-trackers, the goal is unknown. Although they may take up tracking with a specific question in mind, they continue because they believe their numbers hold secrets that they can't afford to ignore, including answers to questions they have not yet thought to ask' (Wolf, 2010).

Figure 4.01—**Mesh 1 (Bicycling mileage 1997–2009), 2010; Acrylic, hardware; 78 × 78 × 20 inches.** Stephen Cartwright

Tiffany Qi's 'College Performance' project is an example: according to the QS site, during her four years of undergrad, Qi tracked her time, organizing and colour-coding it into categories, e.g., 'exercise', 'class', 'games/time-wasting' and 'homework and studies' analysing the data with questions such as:

→ Did Qi's commitment to her studies wane over time?
→ How much did time spent studying matter for Qi's final grade?
→ Did the amount of time spent on fun help or hinder her grades?
→ Did having a job or other job-like responsibilities lower Qi's grades?

The productivist value system behind these research questions is also reinforced in Qi's bar chart visualizations. Prominent colours (red and orange) are reserved for class, studies and exercise, while the 'games/time-wasting' category is rendered in a dull brown. The QS impulse to bare all and let the data 'speak for themselves' (see Chapter 3) not only recalls the Tufte-Shneiderman investment in the notion of data's neutrality, it gives many presentations and accompanying visualizations the air of a public confession (Sharon and Zandbergen, 2017, p. 1704). Aside from tracking diet and exercise, QS-ers have presented visualizations of their sleep patterns, headaches, moods, ovulatory cycles and shopping habits, body scars and their causes. Here, confesses the Quantified Self to her compatriots, is what the data tells you about me: based on this, what can *you* tell me about me?

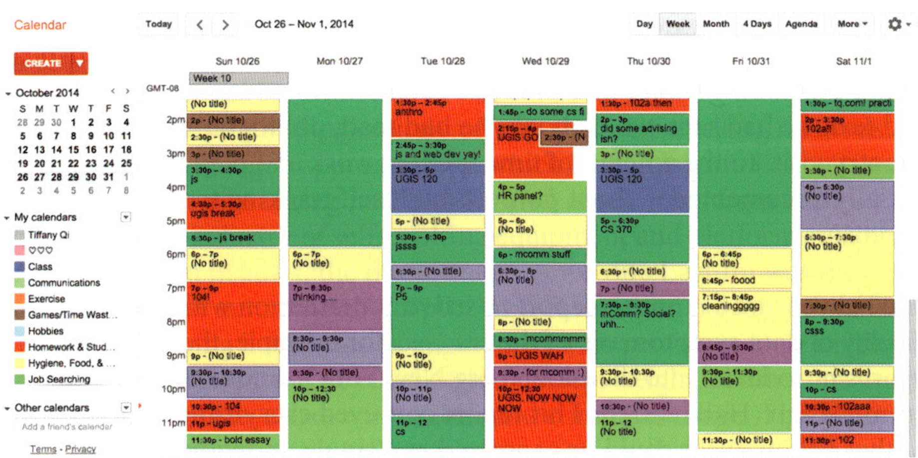

Figure 4.02 — **Weekly calendar.** Tiffany Qi

The sense in which Marey's sequences of images both do and do not present an experience of movement has been the subject of considerable philosophical debate since. Bergson famously invokes Marey's images in several essays of the late nineteenth and early twentieth centuries, in which he is working out the concept of duration (durée), as illustrated in the discrepancy between innate or experienced time (duration) and remembered and recorded time. According to Bergson, the representations of movement in Marey's chronophotographs are illusory in that they are discontinuous fragments that our consciousness spatially fixes: 'Instead of attaching ourselves to the inner becoming of things, we place ourselves outside them in order to recompose their becoming artificially. We take snapshots, as it were, of the passing reality' (Bergson, 1911 cited in Rabinbach, 1992, p. 112).

Figure 4.06—**Chronophotograph: 'Movements in Pole Vaulting' (1885/1895).** Étienne-Jules Marey

For the philosopher Gilles Deleuze, it is through the recomposition of images seen through the inhuman eye of the camera, that seeing is liberated from an organizing viewpoint, and the 'movement-image' becomes possible. In this sense, the chronophotographs present a diagram of the cinematographic function, revealing its radical possibilities. Marey's camera moves as the runner moves, and movement is no longer the synthesis of points in a single line of time, but 'movement itself, in all its diversity, from which single points of view are composed' (Colebrook, 2002, p. 33).

If Marey's visualizations seeded the radical possibilities of cinema, they also expressed and encoded the logic of capitalism. For the capitalist to buy a portion of the labourer's time in order to produce surplus value, 'that time must be measurable and therefore divisible' (Doane, 2002, p. 8). The emergence and internalization of divisible, mechanical time are tracked by historian E.P. Thompson through the early stages of industrialization: he records various conflicts over efforts to standardized time in eighteenth- and nineteenth-century Britain, and the long-standing agrarian and seafaring, task-based systems that were practised in response to the tides and passage of the sun. The day in pre-industrial societies was organized around tasks that needed to be completed within the available daylight, the tides or the 'cattle clock' (depending on the work). Pre-industrial approaches to expressing duration reflected local customs and belief systems:

> Thus in seventeenth-century Chile time was often measured in "credos"; an earthquake was described in 1647 as lasting for the period of two credos; while the cooking-time of an egg could be judged by an Ave Maria said aloud. In Burma in recent times monks rose at daybreak "when there is light enough to see the veins in the hand". The Oxford English Dictionary gives us English examples—"pater noster wyle", "miserere whyle" (1450), and (in the New English Dictionary but not the Oxford English Dictionary) "pissing while"—a somewhat arbitrary measurement. (Thompson, 1967, p. 58)

The impact of 'time discipline' described by Thompson was insidious and far-reaching, extending from bells and clocks to curfews, preachings and schoolings, the suppression of sports, and the division and supervision of labour. 'New labour habits were formed and a new time-discipline was imposed' (Thompson, 1967, p. 90). Pre-industrial labour mixed work and play somewhat indiscriminately, but industrial labour required punctuality; machines set the pace at which people worked, and time became a commodity that could be 'saved', 'spent' or 'wasted' (Adas, 1995, p. 242).

The cultural specificity of time discipline became more pronounced when contrasted with the cultures of the unprepared subjects of the expanding European empires. According to historian Michael Adas, the European explorers and missionaries who went out to Christianize Africa and Asia found the pace of life exasperating. Nineteenth-century observers insisted that Africans had no sense of time beyond that of nature's rhythms, marked by the cycle of the wet and dry seasons, the transitions from day to night, birth to death, or the waxing and waning of the moon (Adas, 1995, p. 244). Because the Africans they encountered had not developed writing, the Europeans assumed they had no history: Edmond Ferry claimed that the peoples of the Sudan had no verb forms to express the past tense (Adas, 1995, p. 244). Adas sums up the European attitude to the so-called primitive colonized cultures: 'Textbooks used in colonial schools urged native school children to reject the slothful ways of their ancestors and embrace the work ethic of their conquerors' (Adas, 1995, p. 258).

Many nineteenth-century writers at the time equated the imperialist projects of Europe with the 'triumph of science and reason over the forces of superstition and ignorance which they perceived to be rampant in the non-industrialised world' (Adas, 1995, p. 204). The control of populations was precisely the challenge of the larger political project of colonization, and time discipline, categorization and enumeration were the empiricist's tools for this. Edward Said's seminal book, *Orientalism* (1978), describes the ideologies that supported the European colonial project of domination and extraction. Orientalism, argues Said, is also 'anatomical and enumerative, to use its vocabulary is to engage in the particularizing and dividing of everything Oriental into manageable parts' (Said, 1978, p. 72). If the Algerian colonial project combined feminization with architectural intervention in an attempt to achieve control of social units, the British project in India was characterized by the generation of a 'vast ocean' of data, counting and measuring land, fields, crops, forests, castes, tribes and so on.

Back in the UK, US and Europe, the work ethic of the conquerors began to reach industrial proportions. The Scientific Management movement, which drew from Marey's chronophotography and its rationalization of the movements of the body, introduced the concept of industrial efficiency in the production lines of the twentieth century. Formal and methodological similarities in the visualizations emerging from Marey and those who followed him two decades later are identified in Siegfried Giedion's influential 1948 history of 'anonymous' design *Mechanization Takes Command*. The 1894 study of a blacksmith wielding a hammer at the anvil, for example, in a multiple exposure photograph subsequently schematized in a drawing by Charles Fremont in Marey's laboratory, predates by twenty years the 'cyclographs' of Frank and Lillian Gilbreth, which sought to reduce the motions involved in a given task by attaching small electric lamps to subjects' fingers and capturing their movements with long exposures (Gainty, 2016).

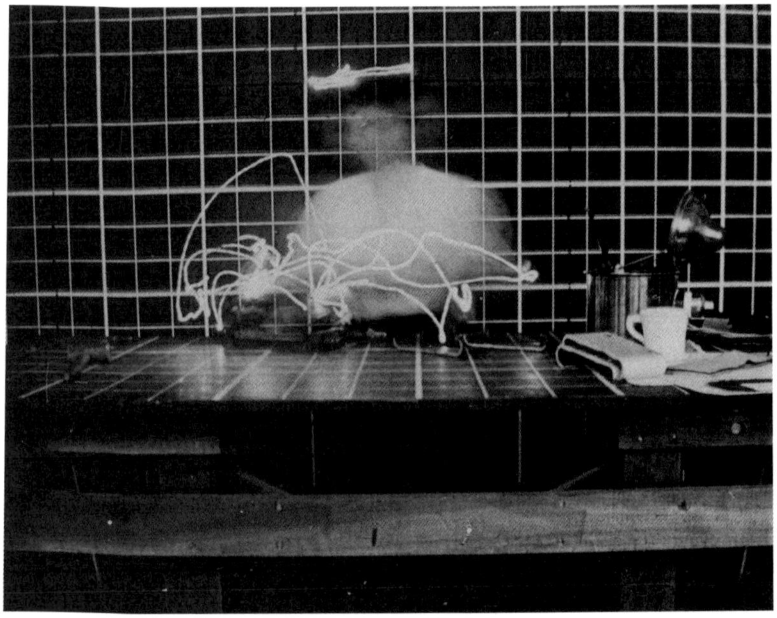

Figure 4.07—**Motion efficiency study by Frank Gilbreth (1914).** Frank and Lillian Gilbreth Collection, Archives Center, National Museum of American History, Smithsonian Institution

Where the Gilbreths used the camera to study human movement and eliminate inefficiency, the mechanical engineer Frederick Winslow Taylor's primary tool was the stopwatch. And where physiologists like Marey were concerned with how a worker's movements might be optimized to reduce fatigue, Taylor's interest was in maximizing productivity regardless of the physiological impact. Taylorism, the name given to distinguish the US-based Efficiency Movement from parallel trends in Europe, was formally established with the publication of Taylor's 1911 *Principles of Scientific Management*. The book describes how 'thousands of stop-watch observations'[7] were made of labourers at the Bethlehem Steel Company in Pennsylvania, to determine how quickly a labourer could push a shovel into a pile of materials and draw it out fully loaded, with different shovel sizes and load weights tried in order to establish maximum efficiency. In another experiment, a bricklayer's motions were reduced from eighteen to five, to increase speed and output. The book effectively handed management a means of controlling the entire production process.

Scientific Management met with fierce resistance, much as the clock and mechanized time brought 'far-reaching conflict' in Thompson's account:

7 https://www.marxists.org/reference/subject/economics/taylor/principles/ch02.htm (accessed 29 May 2021).

zoom out to a landscape of participants and overlay detection rectangles and lines that connect individuals across the room. The addition of terms that appear to classify the relations between apparent strangers, such as 'distant/attraction/interest' suggests that the machine algorithms are able to speculate or detect connections between people—even where there might be none. In the context of an art fair or exhibition, the visualizations naturally prompt amusement, but Lozano-Hemmer explains in an interview that *Zoom Pavilion* is at the same time a 'terrifying' piece, 'because it's the surveillance of relationship' (Lozano-Hemmer et al., 2015, p. 219).

Figure 4.17—***Zoom Pavilion*** **(2014).** Rafael Lozano-Hemmer in collaboration with Krzysztof Wodiczko; photo by Mariana Yañezr

Lozano-Hemmer's work in this sense exposes the limits and biases—and power—inherent in biometric systems. In *Level of Confidence* (2015), which commemorates the forced disappearance of forty-three student teachers from Ayotzinapa, Mexico, a facial recognition system is programmed to search for the students, and when confronted by a visitor to the exhibition, looks for shared physical traits between the students' photographs and the face of the visitor. Its repeated failure to find a match becomes a haunting memorial that also implicates the viewer in the process, challenging the notion of an innocent bystander.

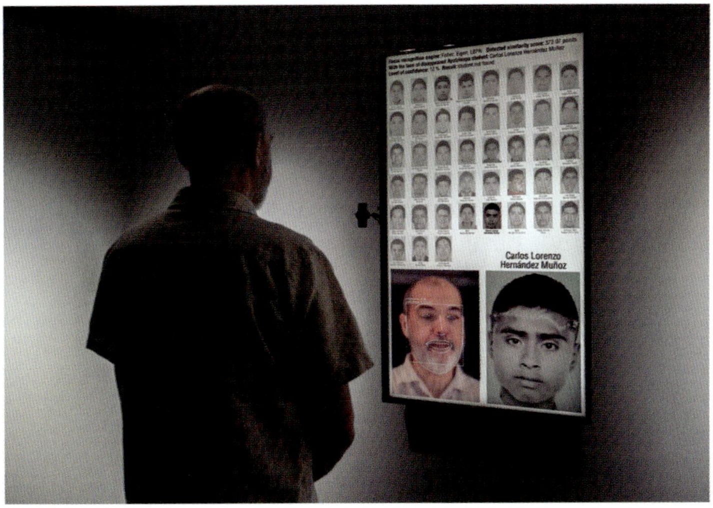

Figure 4.18—*Level of Confidence* (2015). Rafael Lozano-Hemmer; photo by Guy L'Heureux

The possibility of machine error and algorithmic bias in biometric surveillance systems is explored in a study by Shoshana Magnet (2011), which finds that manual labourers often cannot be successfully fingerprinted because their fingers are worn, and that eye recognition readers often fail on people in wheelchairs or people who are too tall, because the iris recognition only works within a range of angles. Simone Browne's extensive study (2015) of the intersection of race and surveillance elaborates on how 'norms' are racialized, as in the example of how street behaviours of white men standing around are coded as normal and the same behaviours of Black men are not. Browne contends that within the field of surveillance studies race remains undertheorized. 'Blackness' argues Browne,

> is ever present in the subjection of black motorists to a disproportionate number of traffic stops (driving while black), stop-and-frisk policing practices that subject black and Latino pedestrians in New York City and other urban spaces to just that, CCTV and urban renewal projects that displace those living in black city spaces, and mass incarceration in the United States where, for example, black men between the ages of twenty and twenty-four are imprisoned at a rate seven times higher than white men of that age group, and the various exclusions and other matters where blackness meets surveillance and then reveals the ongoing racisms of unfinished emancipation. (Browne, 2015, p. 13)

The implications of statistical norms and their negative impact on peoples' lives are felt across the design disciplines, and opportunities to resist them are not limited to the 'satirical and playful practices' of the arts. A visualization project that performs the function of challenging and reinventing the assumptions of early anthropometry is SizeChina, a functional database of head shapes in China. Initiated by product designer Roger Ball while working as a researcher at Hong Kong Polytechnic University, the 3D anthropometric database is the result of fieldwork in six provinces of mainland China to scan the heads of 2,000 Chinese civilians. In the project documentation, extensive photography illustrates the *situatedness* of the data-gathering and the process by which the 3D scans were cleaned up and amalgamated to create a set of anthropometric measurements for designers of eyewear and headgear for Asian markets. The project was initiated by Ball—and funded by the Chinese government—after it became apparent that the 'standard' data set for head shapes was not universal as its status suggested, but based on measurements of 1940s US military personnel: Designers at the US-based Henry Dreyfuss Associates had been tasked with designing equipment for the cockpits of fighter planes, and hired Alvin Tilley in 1946, to draw the famous anthropomorphic charts of Joe and Josephine, 'typical Americans' (Flinchum, 1997, p. 87). Tilley's sources were not exactly clear, but the charts resulted in the publication in 1960 of the book *The Measure of Man*, republished and updated as *The Measure of Man and Woman*—and still used as a reference text by designers today. Ball discovered that both eyeglasses and headgear were ill-suited for Asian head shapes, because of the cultural specificity of the so-called 'universal' data, after a group of Japanese snowboarders explained to him that the helmets he designed for the US company Burton didn't sell in Japan, because they gave a headache to the wearers, who typically had rounder-shaped heads than US military personnel.

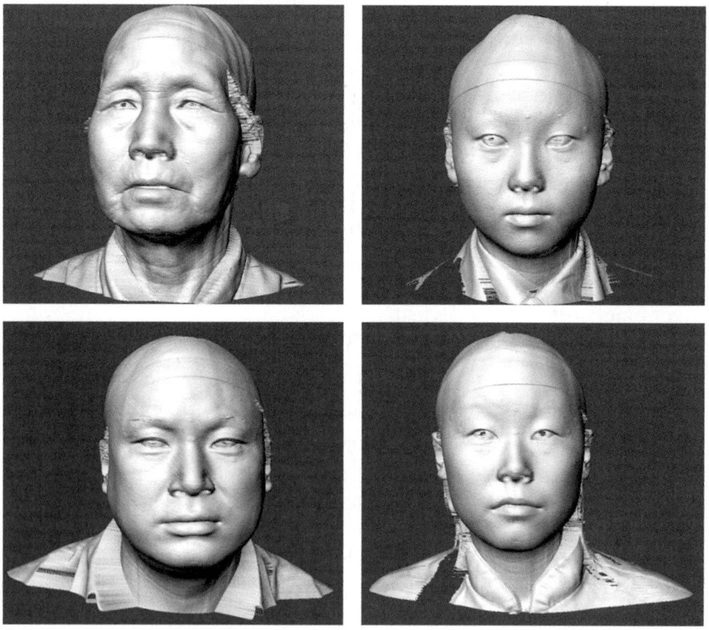

Figure 4.19—**Scans from SizeChina.** Roger Ball

A comparably flawed claim to universality is to be found in the algorithms behind facial recognition technologies, whose failure to recognize race and gender can be traced to the prejudices written into the biometric algorithms (Magnet, 2011, pp. 34–43). Among the repeated failures are miscategorization of sex based on hair length, make-up, expression of emotion (women smile more than men) and clothing. In one experiment, a computer was taught to assign race based on 'component' analysis and two-dimensional vectorized models of photographs, then correlated with a database of manually labelled photographs, which produced a 17.5 per cent failure rate. Magnet finds the racist roots of anthropometry highly problematic, and notes that among those misrecognized by biometric technologies are mixed-race individuals:

> As a result of biometric science, entire communities become what Leigh Star and Geoff Bowker called "boundary objects" Donna Haraway has called "monsters" and Kal Alston has called "unicorns"—uncategorizable objects that do not fit into any classification system. (Magnet, 2011, p. 44)

THE EXAMINED LIFE

Earlier in this chapter, it was suggested, by defenders of the Quantified Self movement, that a more heterogenous ethnography of self-trackers reveals 'meaning-making' through mindfulness and in the communicative and narrative practice of articulating experiences at the boundaries of different domains of knowledge. Sharon and Zandbergen seem to suggest that in the process of quantifying their lives, practitioners achieve a level of self-reflection that is also expressed in conversations and presentations among the community—notably at QS gatherings. Indeed, they find, amid the confessional aspect of the QS presentations of data, an emerging concept of the 'qualified self'—a term used to describe the processes by which quantified data are interpreted and transformed into qualitative narratives that create meaning (Sharon and Zandbergen, 2017, p. 1705).

In closing, we will explore this claim that to quantify oneself contributes to 'knowing thyself', to cite the Socratic aphorism. Much of this chapter has focused on the way in which the 'self' as we know it, is not a raw or pure underlying essence that can be revealed through a lens of numbers, but is constituted by the aims, ambitions, decisions and agendas embedded in the technologies we use and have inherited. Lupton usefully identifies three bodies that constitute the quantified self: the body as machine, the private body ('Not overcoming difficulties becomes firmly positioned as the fault of the private individual rather than of their relative social and economic advantage' (Lupton, 2016, p. 51).) and the spectacularized body.

that is its content' (Sharon and Zandbergen, 2017, p. 1701). In an interview, Frigo describes a process that is analogue and ritualistic. 'I believe I have become more stoic, the love that I put into the curation of my lifelog makes me a happier and more helpful person to others' (Frigo, 2015).

Figure 4.21—Cube installation documenting a 36-year 'Life Stowing' project. Each square corresponds to one of 36 'life works,' made up of 432 hand-welded pixels that can be scanned by visitors. Alberto Frigo

In the serial documentation of one's daily activity, it would seem, is something life affirming, that goes beyond quantification and measurement. This might be better understood in terms of the post-war art, film and literature movements that took as their subject the everyday, such as the French Oulipo group, whose stripped-back literary experiments used as their subject matter the 'most obvious, most common, most colourless' (Perec, 1974/1997, p. 50), or the artist On Kawara, who for eleven years sent friends a daily postcard recording the time he awoke, his geographical location and details of the people he had met, places he had visited and books he had read.

Figure 4.22—*Signs of Life* (1994). Alfredo Jaar.
Courtesy of the artist, New York

The ritualistic aspects of self-tracking begin to suggest a very different notion of self-hood when situated in close proximity to death and trauma. The artist Alfredo Jaar took inspiration from Kawara to make his *Signs of Life* postcard project after visiting Rwanda following the 100-day genocide in 1994, which led to the deaths of one million Rwandans. In August of that year, Jaar travelled to Rwanda to bear witness to the aftermath of the genocide, which has been called the 'failure of humanity' (Dallaire and Beardsley, 2004)—the failure of the international community to respond, its absence in Western media, and the result of systemic racism.

Jaar's Rwanda Project, created between 1994 and 2000, comprises twenty-five works including *Signs of Life*, the postcards Jaar found in a destroyed post office in Kigali, and mailed, each with the name of a person he had encountered during his visit and the handwritten message '___ is still alive!' (Duncombe and Lambert, 2018). The project, described as both 'a dirge and an exaltation' (Gervat, 2019, para. 6), was followed by *Searching for Africa in* LIFE, a visualization made up of all 2,128 covers of the US photographic news magazine LIFE, published between 1936 and 1996 (see Figure 1.13). Notably absent from the covers is any evidence of coverage of human life and culture on the African continent, revealing a blind spot in the magazine's supposed world view.

The significance of the everyday in the arts, politics and philosophy has been extensively discussed, but its relevance to data visualization sheds a useful light on the Quantified Self movement. If self-tracking at one level suggests a desperately techno-utopian bid to enact an individualistic productivism that plays into the hands of a biometric surveillance society, at another level it holds the potential to re-enact the selfhood that refuses to be commodified, surveilled and homogenized.

Amoore's investment in 'satirical and playful' practices in the arts has a grounding in the work of Henri Lefebvre, among other theorists to have focused on the politics of the everyday, particularly in the context of the post-war years. In his *Critique of Everyday Life*, Lefebvre argues that art, like play, is 'transfunctional—it is both useless and has many uses, acting as a kind of "play-generating yeast" in the everyday, fermenting, agitating and disrupting' (Lefebvre, 1961/2008, p. 14). A visualization practice that is able to ferment, agitate and disrupt is quite a different proposition from the positivist mantra of the QS life-tracker, but perhaps closer to what Sharon and Zandbergen were leaning towards in their case for meaning-making through narrative practices of articulating experiences at the boundaries of knowledge domains. In its distrust of the heroic and spectacular, Lefebvre's account of the everyday confronts 'the bureaucracy of controlled consumption' (Harris and Berke, 1997, p. 3). There is no better rejoinder to the charge that self-tracking is only a practice of self-commodification and self-surveillance. ■

people living in a city—'the signature of humanity'. But Ratti argues that for big data to have a big impact it needs to be 'analysed and digested':

> Today, that is done by visualization experts, who might make an image, a graph or a video; what we are now working on at the lab is making a tool that puts data at your fingertips ... When everyone—from scientists to politicians to everyday commuters—has the opportunity to explore invisible dimensions of the city around them, we will see radical behavioural change. When the city is revealed, particularly in real time, we all will become smart citizens. (Ratti and Claudel, 2018)

The implication here is that with the forward march of sensing and visualization technologies, the smart city will arrive at a near-synchronous state of monitoring-visualizing-tuning. A hurricane is forecast, evacuation plans are e-distributed among the populus and the flood barriers automatically rise; air quality approaches a dangerous level in the city centre and traffic is re-routed. Speculative visualization projects such as Real Time Rome, or Arup's Melbourne Smart City (Arup, 2010)—which hypothesized a floating 'net' of data to communicate patterns of urban activity (Moere and Hill, 2012, p. 34)—seem to suggest that the feedback loop between the technology and its citizens will be increasingly streamlined to the point where humans are fed a stream of just-in-time, accessible information (via floating data nets or embedded displays), or cut out of the loop completely.

However, such envisioning tends to glide over the point that, as Orit Halpern puts it, a visualization is 'non-synchronic' with the event it is depicting and translating. Between the event and its visual representation (even if it is animated in 'real-time') is a lag:

> This non-synchronicity preoccupies our imaginings of "real-time" interactivity and data visualization, driving a constant redefinition of the temporal lags between collecting, analysing, displaying and using interfaces. (Halpern, 2014, p. 22)

Arguing that vision and reason have become intertwined with governmentality, Halpern introduces the case study of the South Korean smart city Songdo, which is being built from the ground up on drained land from the Yellow Sea south of Seoul. A collaboration between the South Korean government with US property developer Gale International, architect Kohn Pedersen Fox and network infrastructure provider Cisco Systems, Songdo is conceived as a 'testing grounds for the future of human habitation', and part of a free economic zone integrating an airport, infrastructure, technology and housing (Halpern, 2014, p. 2). Songdo's eco-friendly techno-infrastructure includes rainwater collection, remote heating, ventilation and air conditioning control, and waste disposal via pneumatic tubes to an underground waste facility. Flow diagrams on KPF's website are used to contrast Songdo infrastructure with conventional systems (KPF, n.d.; Lobo, 2014).

The assumption that 'life itself' can be managed by bandwidth, and that data presents 'stability, wealth and sensorial pleasure' is built into the foundations of Songdo, which is constructed over an infrastructure of tunnels for fibre optic cabling, and comprises networked buildings whose surfaces function as interface, wired with sensors to track behaviours and transfer information. According to Halpern, Songdo represents a current obsession in urban planning for cities made to hold millions of people, cities that are sold by engineers and consultants as 'commodities' built on a 'biopolitical' hypothesis that data can be gathered and traded in 'recombinable units of attention, behaviour and credit' (Halpern, 2014, p. 5).

Where the machine-age city offered the promise of frictionless efficiency and precision in the movement of goods, the twenty-first-century smart city promises security through predictive policing and smart clouds, with zero waste helping the smart city 'do its bit' for the environment. As part of an ongoing Google Earth Outreach special project on air quality in cities, air pollution sensors have been installed on selected Google Street View vehicles to produce data for a map of Copenhagen, developed in partnership with the City of Copenhagen and Utrecht University. The map shows the street level concentration of black carbon and ultrafine particle pollution, which, according to Google's promotional material, is being used by the city to work with architects and designers 'to rethink the city for the future' (Powell, 2019).

Again, details on how Copenhagen's officials, designers and Google are 'rethinking' the city are not forthcoming. This may be simply because it is too early to spell out how monitoring and visualizing air quality might lead to steps to improve it, but our point is to show how the rhetoric of smart city visualization tends to lean rather heavily on the notion that *visualizing* the problem is tantamount to *addressing* the problem.

The same critique might be applied to any number of large-scale visualization projects in which important environmental information is represented continuously on large-scale surfaces. A recent project by RNDR in the Hague indicates that there is a significant amount of work for designers in making invisible flows visible through experiential visualization. The tunnel over the A2 highway in Utrecht (part of the urban development master plan for Leidsche Rijn by Maxwan which by 2030, will bring 110,000 people to live in the area) makes the district into a single entity, hiding the cars on the motorway, and keeping their noise and exhaust fumes, underground. RNDR's light installation OPEN HIGHWAY—14 light bars (11 m long) placed to correspond with the lanes below—makes the hidden highway visible again and 'shows that you can make something beautiful out of something intrusive' (RNDR, 2019).

Figure 5.01 — **Open Highway installation by RNDR.** RNDR — Jeroen Barendse (traffic flows) and Juri Hiensch (installation)

Beautifying data is a function of visualization, discussed in Chapter 6. Making it also actionable is more central to the question of this chapter — how might a critical visualization respond to the challenges of the contemporary city? In light of the catastrophic bush fires in Australia, such proposals as Arup's data 'net' could take on the potential role of emergency signage rather than a gentle nudge to 'look after the earth'.

As cities and megacities begin to take steps to protect themselves against anticipated extreme weather events as a result of climate change, visualization also plays an important role in explaining and persuading various publics that proposed planning projects are in the interests of stakeholders. An instructive example is the array of renderings, maps and visualizations presented for New York's $930 million Rebuild by Design project by Bjarke Ingels Group (BIG) and Dutch firm One Architecture & Urbanism. The images depict the plan for a ten-mile-long flood barrier to protect areas at the low-lying southern tip of Manhattan from floods and storm surges, described by architect Bjarke Ingels, founder, and creative partner of BIG, as a series of public spaces and viewpoints, or 'a string of pearls of social and environmental amenities tailored to their specific neighbourhoods, which also happens to shield their hinterlands from flooding' (Gilbert, 2014). The scheme, named eponymously (with a rhetorical finger beckoning its publics) 'BIG U', is strategically conceived not as a barrier but as a berm, with renderings of future citizens walking, sightseeing and even wading through flooded terraces planted with salt-water-tolerant flora. To make such a scheme 'urbanistically successful' Ingels told the design blog *Dezeen*, 'It needs to happen rooted in a dialogue with the different communities' (Howarth, 2018).

Figure 5.02—**The Rebuild by Design project, a $930 million flood barrier plan for lower Manhattan.** Bjarke Ingels Group (BIG) and Dutch firm One Architecture & Urbanism

Unlike New York's grand planning projects of the twentieth century, Rebuild by Design was premised on a promise of early-stage community consultation. But which communities? In its striking maps of sea currents, flooded areas and renderings of resilient New Yorkers calmly participating in a kind of flood-friendly blue-skied utopia, BIG U adopts a surprisingly parochial position vis-à-vis the larger context of another storm surge like that experienced with Hurricane Sandy in 2012. In his 2017 critique of the project in the book *Extreme Cities*, Ashley Dawson cites a report by a climate scientist who argues that the flood barrier increases rather than decreases risk, providing a false sense of security (like the New Orleans levees) and displacing danger into other areas:

> Where will the water that the BIG U turns aside go? It is likely to end up in adjacent communities with large poor populations such as Red Hook, where hurricane Sandy hit public housing particularly hard. Interestingly, BIG looked at flood defences for Red Hook in its research, but this was not part of the final proposal. It should not be particularly surprising that defense of Wall Street garnered more attention and funding than defense of a historically poor (if rapidly gentrifying) neighbourhood such as Red Hook. (Dawson, 2017, p. 160)

Urban planning for extreme weather, it seems, is being performed with the promise of resilience among a small, fortunate elite—in this case, those living and working in lower Manhattan, an area which is home to the world's two largest stock exchanges, and whose real estate is among the most expensive in the world. As has become quite common in contemporary geo- and environmental politics, the sense of a shared world that dominated visualizations of the last century—for example, Buckminster Fuller's 'geoscope' proposal for a 60-metre globe lined with screens graphically representing an inventory of the earth's resources and developmental trends—has been supplanted by images of model city quarters for a surviving few (Wigley, 1997). Latour characterizes a position of 'post-politics' that seems to account for the impulse to stake out and protect a defined gated community whose future is, by definition, short term, since all territory is now threatened by

This picture is elaborated by the anthropologist Anna Lowenhaupt Tsing's evocation of the precarious livelihoods and environments of twenty-first-century capitalism thwarted by ecological catastrophe: in place of a dream of continuous modernization and progress is a picture of 'patchy' capitalism in which 'the concentration of wealth is possible because value produced in unplanned patches is appropriated for capital' (Tsing, 2015, p. 5). Tsing's book, subtitled 'the possibility of life in capitalist ruins' follows the trail of the matsutake mushroom, which grows in human-disturbed forests and is said to have been the first living thing to grow in the blasted landscape of Hiroshima.

In light of patchy capitalism and a situated mapper, it seems pertinent to consider in closing how data visualization is providing support for desirable shifts in behaviour. In the simplest terms, public displays of real-time data are an increasingly common means of representing statistical trends to urban citizens, and, in some cases, a trigger to escalate those trends. Nuage Vert by artist group HeHe, the winner of Ars Electronica in 2008, *détourned* the cloud emerging from a chimney at an electricity plant, turning it into a projection surface. As energy was saved by city residents, larger areas of the cloud were lit up by green laser light, functioning as a motivational visualization. Arguably, more actionable is the bicycle barometer, a real-time visualization that clocks how many cyclists have passed the sign that day and year, launched in Odense, Denmark in 1999, and is now installed in cities around the world. Its function is simple: to motivate more people to use bicycles (knowing that they are not alone) and to equip policymakers with data to inform decisions around cycling infrastructure. In effect, the bicycle barometer uses the 'nudge' theory of behavioural design documented and advocated by Thaler and Sunstein (2009), which became the basis of a British Government unit in 2010.

In their survey of practices in the public visualization of urban data, a useful distinction is made by Andrew van de Moere and Dan Hill (2012), between three aspects of urban visualization: *functional* visualizations that serve to provide a 'trustworthy and persuasive' experience to a large audience, *informative* visualizations that create a direct feedback loop between the actions of city dwellers and their environment, and *situated* visualizations that take into account the characteristics of their physical location, reflecting on relevant local issues. The bicycle barometer, curiously, satisfies all three of Moere and Hill's criteria, in providing trustworthy local information that creates a direct feedback loop.

Figure 5.09—**For Nuage Vert, emissions from a Helsinki coal-burning power plant (2008) and a Paris incinerator (2010) were illuminated with a high-power green laser animation.** HeHe/Antti Aahonen

In conclusion, it would seem that the missing ingredient in smart city rhetoric, is one that takes on board the topic of everyday life as a site of potential political transformation, as articulated by Henri Lefebvre:

> ... it is in everyday life and starting from everyday life that genuine creations are achieved, those creations which produce the human. (1961/2008, p. 31)

In his book *Down to Earth*, Bruno Latour (2018) argues for a redefinition of politics as what leads toward the earth rather than the global or national. Ecological degradation on a planetary scale has become increasingly visible, he writes, its effects visible in three simultaneous political phenomena: vertiginous escalation of inequalities, massive deregulation, and climate change denial. One effect of ecological degradation is that the historical colonizer and the colonized are thrown together in something like a new, bizarre bond, replacing the classic conflict. The colonizer asks:

> "How have you managed to resist and survive? It would be good if we too could learn this from you." Following the questions comes a muffled, ironic response: "Welcome to the club!" (Latour, 2018, p. 23)

Historically, colonizers have often depended on Indigenous peoples for survival, only to disavow their dependence in order to displace them. But Latour's belief that the climate emergency casts a new light on this relationship underlines exactly why decolonizing design is both an imperative and awkward, humbling project for Westerners, as we try to turn our critical tools on ourselves, embrace a praxis-based learning and invite those we have previously excluded, to join us in the project. The distrust and weariness at our relentless narcissism can only be countered with the reminder that we're all in this together. ■

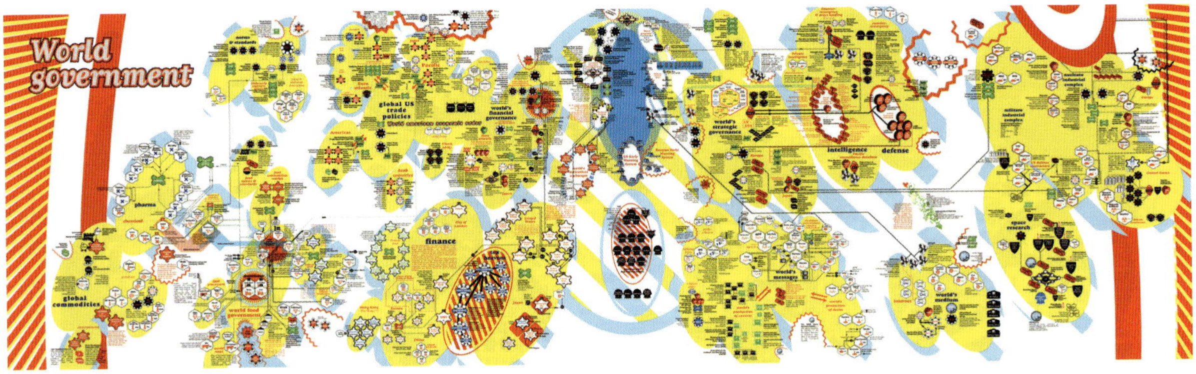

Figure 6.03—**Gouvernement mondial (2013).** Bureau d'études

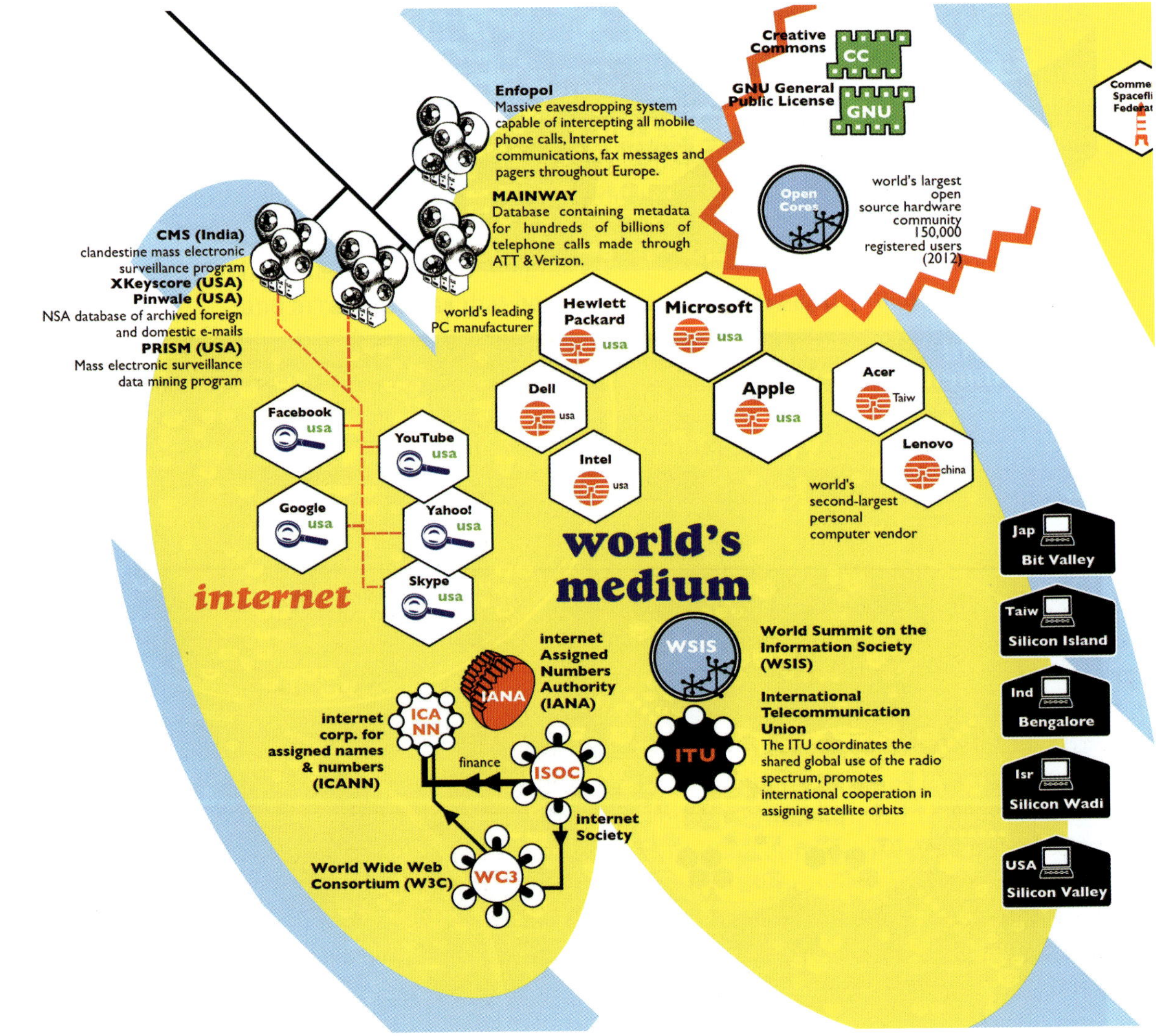

Figure 6.04—**Detail of Gouvernement mondial (2013).** Bureau d'études

Beyond Aesthetics and Representation

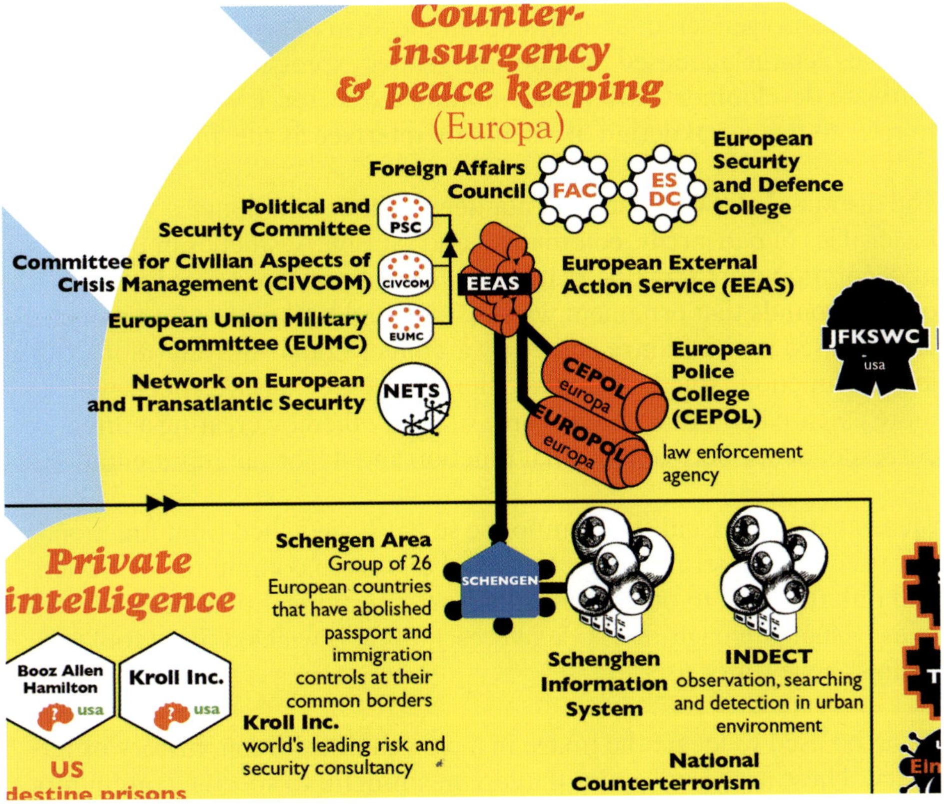

Figure 6.05 — **Detail of Gouvernement mondial (2013).** Bureau d'études

It has become a truism in recent years that beauty is contingent on the viewer's own context, priorities, tastes and cultural backgrounds. As Umberto Eco notes, 'The mass media no longer present any unified model, any singular idea of beauty' (2004, p. 426). This aligns with the dichotomy where subjectivity is valued less than objectivity—where the former is variable and deceptive, and the latter is constant and true.

In their panoramas, Beehive Design Collective uses intricately hand-drawn figures to tell their story. These are meant to be pleasurable for the viewers while the stories told are of struggle, resistance and violent oppression. There is a beauty to the story maps that helps engage the audience in much the same way as Mexican muralists (e.g. Rivera, Orozco, Siquieros) use composition, scale, colour and figurative drawing in complex panoramic murals to address people in different spaces (i.e. pedestrians, visitors, workers). Aesthetics as beauty are necessary dimensions of experience. As bell hooks (2007) reminds us, aesthetics and beauty bring pleasure and are necessary for surviving poverty and hard living.

Aesthetics is also frequently disparaged as being solely a matter of subjectivity; taste involving the use of style for making form or decorating a surface. It is also gendered, as aesthetics in the form of beauty products and services is largely focused on feminine gender expressions. In fields such as software development, still a male-dominated sector, it was once popular to refer to the visual design of a software interface as the 'lipstick'.

The dismissal of the aesthetic as not deep, real or functional is supported by legacies of patriarchy, colonialism and racism. Aesthetics in the form of decoration were repudiated by Adolf Loos (1908/2002), for instance, on the grounds that ornament was wasteful and backwards. Functionalism, as expressed in the mantra coined by architect Louis Sullivan (1896), 'form follows function', sought to replace the language of surface effects with a more essential and pragmatic source of utility thereby creating a direct correspondence between original function and its formal representation.

Yet aesthetics as visual style cannot be so easily detached from function. It is used more instrumentally to appeal to a given community of viewers and link the work to other forms that carry meaning. Sometimes referred to as 'visual language', the style of presentation is wielded like a tool to address a viewer or user.

It can be used to locate the image in a particular tradition, history or territory. These are visual codes that are meaningful to specific communities of people and require knowledge to understand. We may connect very deeply to an image because of this shared understanding between maker and viewer.

Drawing style can also be used to make the data representations more approachable to a viewer. It can remind the viewer that the map or visualization was human-made and therefore subject to bias or error. For instance, J. Wood et al. (2012) found that participants in an experiment were more willing to engage with and critique visualizations that had a hand-sketched style rather than a clear line style.

Figure 6.06—**Bar chart displayed using the Handyrenderer for Processing. J. Wood et al. (2012) found that participants in an experiment were more willing to engage with and critique visualizations that had a hand-sketched style, rather than a clear line style.** Wood et al.

Beyond Aesthetics and Representation

Figure 6.07—**Map showing bicycle availability using Handyrenderer, J. Wood et al (2012).** Wood et al.

Visual style can therefore communicate certainty, or in some cases, authority. Clean lines and a minimalist aesthetic may look more scientific and objective. The modern minimalist aesthetic epitomizes a modern attachment to universalism and function devoid of culturally specific ornamentation. This coincides with a communication model (formalized later by Shannon and Weaver, 1964) that prioritized the integrity of the message and aimed to reduce noise in the signal. Elements that are extraneous to the essence of the message are seen as either noise that makes the message less clear or, worse, communicates its own cultural specificity which runs counter to a virtue of universal accessibility. This was taken up within cartography research in the first half of the twentieth century as a scientific way of improving cartographic communication; that is, the efficient transmission of knowledge through a map medium.

Cartography communication science (MacEachren, 2004) had great appeal to many in the field, as it promised an objective and testable way of increasing the efficiency of spatial data representation—an advantage to planning and analysis for a variety of potential users (e.g. government, industry, science and military).

Cartographers employed a minimalist visual language and the simplicity of lines and geometry to lend an air of objectivity, universality and clarity. The sparse treatment of visuals suggests a more direct correspondence between the data and the representation, and less human involvement. It communicates that the image has been reduced to its bare minimum. It has been polished through successive passes to remove the unnecessary and the contingent. And in doing so, it indicates something essential and closer to a transcendent type, or perhaps an ideal.

for a cartographer. John Pickles notes how a map makes a 'representational claim [that] was always predicated on making the many specific decisions, compromises, adjustments, generalizations, and slippages that characterize all knowledge production disappear. That is, it erases the work the map actually does' (Pickles, 2006, p. 349). It is the clean aesthetic of maps that intensifies the erasure of the messy traces of human work and judgement performed. Framed in the influential terms of Shannon & Weaver as a communication process, the visualization should represent the concept or number (i.e. clarity of signal), rather than display the residue of its construction (i.e. noise in the channel).

AESTHETICS AND PERCEPTION

Visual representation is the translation and amplification of how a site is made legible to a viewer or user. We represent the data in visual form in order to help our eyes make connections and see patterns. As information visualization researcher Ben Shneiderman (1996) states, 'The eyes have it'—the eyes make great use of visual form to help gain insights. As such, this form of representation also affords an aesthetic experience—one in which the body, through the eyes, senses shapes, distances, colours and patterns.

Beyond Gestalt principles of visual perception subsequent psychophysics research has discovered how *preattentive processing* works to helps us distinguish between different visual stimuli (Ware, 2013). The experience of preattentive processing is that of immediate recognition of a spatial arrangement of visual elements (i.e. the grouping of similar objects, connecting nodes in a network). This is used as a building block for designing visualizations and it is in this way that Shneiderman claims that the eyes do much work in finding shapes within data visualization.

Framed as a support for thinking, visualization research has largely focused on the technical and cognitive aspects of the mental tasks of analysing data by visual means. Work in the form of *cognitive load* is understood as a quantity that needs to be reduced so that cognition increases in capacity, speed and efficiency.

Take for instance the 'Visual Information-Seeking Mantra', which describes an optimum way of interacting with data: 'overview first, zoom and filter, then details on demand' (Shneiderman, 1996). This is a useful rule-of-thumb when thinking about how a user might best work with a data visualization with an interactive interface. Managing the appropriate amount of relevant information is at the heart of this design process. What does the user need in order to find the information or the pattern in the data? How does the user understand details within the context of the large data set? How can the user easily compare different data points? How does the design minimize irrelevant data (i.e. noise) and maximize relevant data (i.e. signal)?

The focus on developing a more efficient interaction between the user and a visualization echoes the communication paradigm of cartographic science discussed earlier. Developing better ways of representing data types and structures so that humans can search and find information has been a core focus of much scholarly research in information visualization. What is often left out is the context and meaning generated at the site of production and the site of reception of a visualization where the sender and receiver are located.

It is useful to note that MacEachren, a one-time proponent of a *cartography-as-communication* paradigm, has since critiqued the narrow focus of this approach and has instead argued for a mixed method that acknowledges the technical precision and potential advancement in design of maps with the social, political, cultural aspects of maps that prioritize interpretation and context (MacEachren, 2004, p. 12). To advocate for a mixed method, MacEachren makes an important distinction between communication and representation. Where communication is a technical problem between a sender (cartographer) and receiver (user) through a medium (map), a representation is a non-linear process in which meaning is constructed through an interaction among components such as designer intention, map, map context and user need.

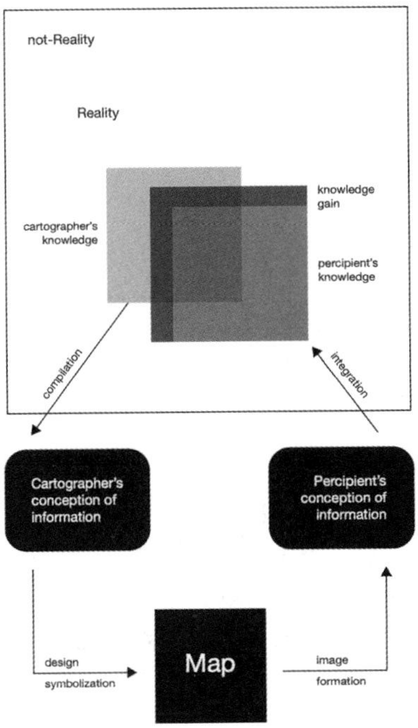

Figure 6.09—**Alan MacEachren's 1979 view of cartography as a process of graphic communication.** Redrawn from MacEachren's 2004 book *How Maps Work*

Representation takes place through translation. An example would be the translation of a living thing into an image of the thing. If we think of a territory translated into a map, we can see that the endless complexity of a living system can be translated into a static two-dimensional image. These translations have a great ability to be copied, distributed, taken apart and reassembled—all while having a claim to representing the original territory. Latour developed the term *immutable mobiles* to describe the documents, maps, illustrations, charts, etc. that point to a thing in the world (e.g. the contours of an island, the temperatures recorded during a scientific experiment). The data and the accompanying visual representations are mobile in that they can be shared widely and at relatively low cost. They can also be collected and assembled into larger documents that stake a claim of knowledge over a thing, a place, an event or a people.

In visualization, representation starts not in the visual artefact but in the counting and measuring. In other words, quantification is representation (Stray, 2016). It is a process designed to select part of a complex field of interacting phenomena and translate it to a stable and computable form. In other words, before there is *visual* representation there is *conceptual* representation.

If we extend the range of meanings of *representation* we can also see its political dimension. Representation is also a form of administration when the things counted are populations of living beings. Just as in political representation, where someone is a representative of another person or group of people, representation in data visualization or mapping is a stand-in or proxy for the thing we are inspecting.

James Corner (1999) describes a double-sided character of mapping in which there is an analogous relationship to physical space and an abstract (or conceptual) relationship. This is the power of maps to both allow us to point at a feature that feels naturally to be the place we may inhabit while also allowing a play of concepts that far outstrips this physical reality—deep history, stories, people and connections to other territories. The bridge between aesthetics and representation reaches between what is sensible and what/who is made available to the senses through visualization.

Given that historically, mapping, visualization and quantification have so often been used for the purpose of command and control, what do critical practitioners do with this legacy? Some play with the utility of the map (e.g. Jorn and Debord). Others use critical maps to combat dominant maps and dismiss essentialist ideas of the map as a tool of domination (e.g. Bureau d'études, Philippe Rekacewicz). Still, others develop a critical practice in visualization which acknowledges that the objects created

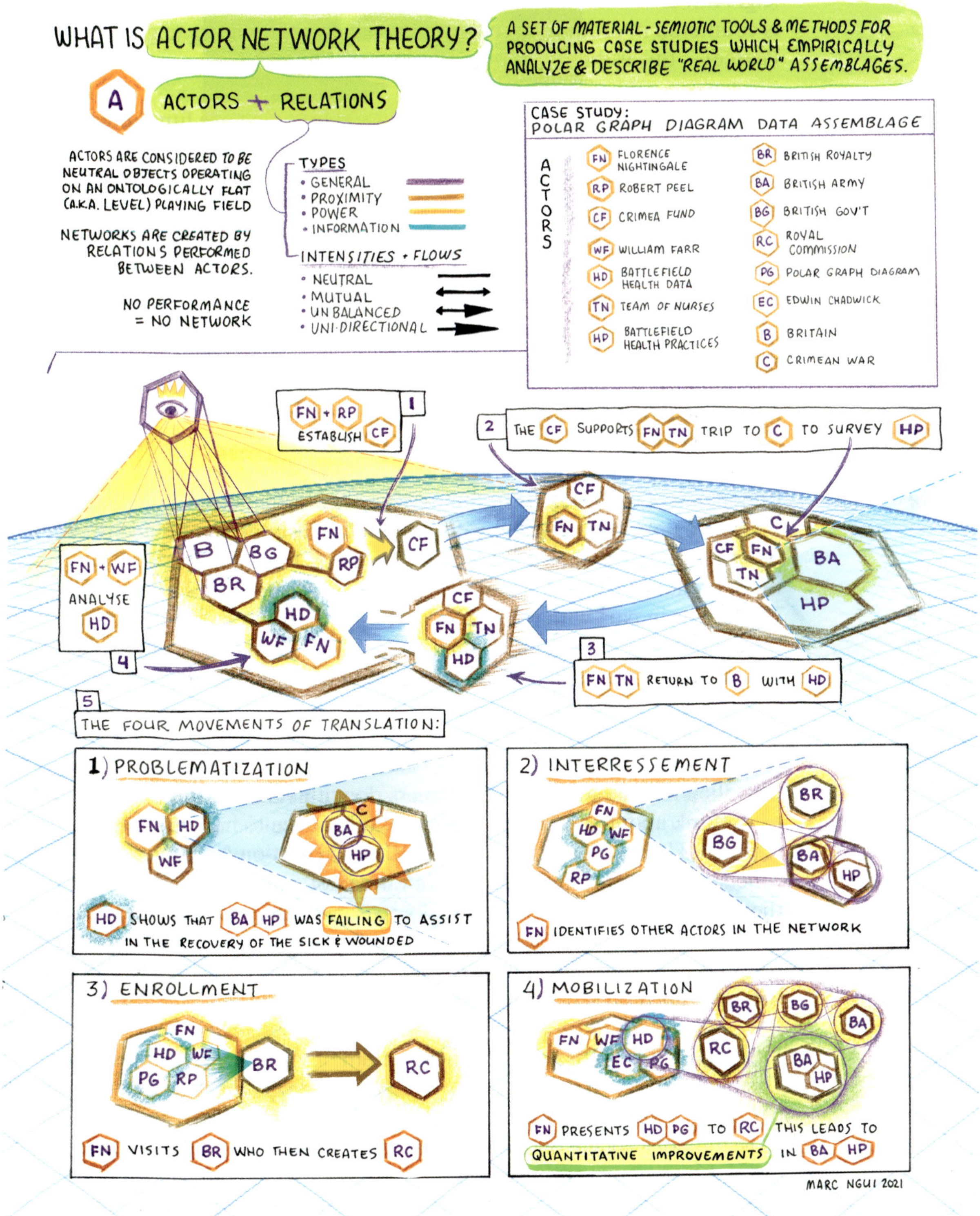

Figure 6.10—ANT. Marc Ngui (see p.61)

Iconoclasistas, while dedicating a good amount of effort to the production of comprehensive maps (e.g. critical mappa mundi), direct their focus to aspects of community-led mapping. In this way their aesthetics are concerned with the representational and the relational. We also see this dynamic in Beehive Collective's projects which are produced over a long period of time, often with residents from the areas they are representing directly involved in the production (e.g. Detroit). They are narrative and explanatory diagrams, which don't use geometric forms and labels but figurative drawings. They are designed for an interlocutor to work with audiences for the visualizations to come to life. They are not fully self-sufficient forms of communication, but rather tools to think collaboratively. Representation thus takes a turn towards performance and affect. Beehive Collective creates communities of production and presentation for the work where the narrative map is inextricably attached to the process of visualizing.

Through the work of these collectives, participation in the creation of visualizations is a form of political representation. They embody an ethos of 'nothing about us without us', where there is a greater correspondence between representations of a community and/or territory and the people who are part of it. For instance, Iconoclasistas' visualization offers a great example of inclusion and participation for the purposes of representing knowledge of specific neighbourhoods.

One of the reasons why groups like Iconoclasistas and Beehive Collective created visualization projects for communities is to combat dominant stories that continue to marginalize these same communities. They offer hidden histories and situate data within radical frameworks (e.g. feminism, indigeneity, resistance).

Cultural theorist Stuart Hall developed a useful framework for understanding representation in media and popular culture that can be used to think about how we consume dominant images about ourselves. This can also be extended to think about visualizations and maps. For instance, Hall notes that representation has a notion of re-presenting something that was already there—as if it's just presented again and therefore is the thing in itself (in fact, he argues, the process of representation itself constitutes the world it aims to represent). This is similar to how data also has a notion of being raw and as being an index of something already there—as discussed earlier in this book. Yet, representation occurs when meaning is given to that which is depicted either as contextualization or framing. In other words, both images and data have the effect of structuring the world they re-present (S. Hall, 1997).

Beyond Aesthetics and Representation

Hall explains that the 'event' has no meaning until it has been represented, therefore when we understand an event, representation has become an integral part of it. We can reapply this notion to think of data as having no meaning by itself until it is represented, framed, visualized, interpreted. Representation is often seen as a dichotomy between accurate or distorted (i.e. true or false), but it is more complex and better understood as a creative-active process.

We can also contrast this notion of representation with the communication model used in mapping and visualization. While in a simplistic communication model there is a focus on the correspondence between the original message and the received message, in representation, as elaborated by Hall, a message is necessarily interpreted in many ways and gains meaning not via the source but through the context of reception. The importance of Hall's insight is that the process of meaning-making takes place through other cultural contexts and ideologies that impose structures of meaning that reproduce racism, sexism, homophobia, amongst other forms of oppression. It is not only the denotative content of an image (e.g. this is an image of person) but also the connotative level (e.g. this is an image of a hero).

We can use Hall's concept to analyse the trajectory of visualization work done by W.E.B. Du Bois, for instance. The maps and visualizations prepared by Du Bois and his team for the Paris Exposition in 1900 sought to fix the interpretation of what Black life is like in the United States for the audience attending the international event. The charts displayed as part of the American Pavilion at the exposition were intended to clearly show the world the progress of the African American standard of living after the civil war, as well as the high level of economic output and cultural production. Du Bois states that he sought 'to show: (a) The history of the American Negro. (b) His present condition. (c) His education. (d) His literature' (Du Bois, 1900). This exposition worked to combat white supremacist stereotypes, conceptions of difference and ongoing dehumanization of non-white people.

Du Bois was concerned with using data and representing facts to address conceptions of life in the USA. Du Bois, in his essay reflecting on the exposition, laments the poor state of scientific rigour in the description of sociological facts and therefore presents his own research to demonstrate methodological precision (Du Bois, 1900). This concern for the denotative accuracy in his charts and maps was coupled with his concern for the connotative power of his exposition. Du Bois displayed data gathered and visualized by himself and his collaborators, along with a collection of portraits of Black life, books and periodicals. In this way, his work was very innovative. In the presentation of demographic data as a form of public education, it predates the work of Isotype. Du Bois contextualized for a decolonial praxis. It has, in fact, 'failed to account for the emancipatory struggles of our time, in part at least because it shares with the bourgeois thinking it criticizes the same epistemological foundations that suppress the cognitive dimension of social injustice, and thus renders universal the Western understanding and transformation of the world. Moreover, it sees itself as a vanguard theory that excels in knowing about, explaining and guiding rather than knowing with, understanding, facilitating, sharing and walking alongside' (Santos, 2016, p. ix).

As visualization is a process that embeds data and representation within cognitive and cultural processes, we take Santos' challenge to heart. One of the dimensions we've chosen to focus on in this book is the work of making and maintaining relations within the process of visualization. This aligns with Santos' working *with* and sharing in a struggle. It prompts us to find a range of ways of making futures that de-prioritize Western frameworks—even critical traditions—which Santos and others are articulating through notions such as *epistemologies of the south*, *decolonization* and *pluriversal thinking*. This, we believe, is where the future work of critical visualization practice must be directed.

One of the reasons we think this is needed is due to the continuation and extension of colonization by means lauded as progress. In the past decade, we have seen a massive increase in the collection of data from users of internet media (e.g. video/photo sharing, online shopping, social media, video/music streaming). This is often accepted as part of the deal, i.e. that data is collected as 'payment' for free searching, navigating, mapping and so on, but it also amounts to a form of data colonization (Couldry and Mejias, 2019) that is reproducing and intensifying power asymmetries by extracting value from work and leisure.

Extraction is a key logic within colonization and capitalism. Historically, this has taken the shape of extending empires from a centre of power (e.g. Britain, France, Spain, Belgium), by subjugating through military violence and cultural assimilation territories and peoples valued as resources (i.e. raw materials and labour). More recently, this has taken the shape of continued settler colonialism, which has further dispossessed Indigenous communities from their own territories and ways of life (e.g. Canada, USA, Australia, Brazil). This works in tandem with the transformation of global relations between rich countries and poor countries where asymmetric trade agreements (as well as military intervention, when necessary) are used as a tool for continuing the extraction of value.

Extraction, rather than sharing with and walking alongside, is a ubiquitous logic that permeates so much of what we do in the everyday. We've described how this logic is also present in the making of data. Data is so often seen as a resource that can be converted into insight both for good and for profit. The invocation that data is the new oil, as mentioned in Chapter 3, further reminds us of the inherent logics of appropriating, amassing, abstracting, transforming and commodifying.

Macarena Gómez-Barris (2017) notes that to resist extractive logics we must recover a basic human faculty. This faculty is part of her own work—she states that, 'by making visible micro spaces of interaction and encounter within geographies where coloniality has left and continues to leave a deep imprint, [...] renewed perception offers a method for decolonized study' (p. 2). It is important to acknowledge that Gómez-Barris focuses on parts of South America where colonization has officially ended but extractive power relations, as stated above, continue. These are submerged and non-disciplinary perspectives that are situated and 'saturated in coloniality' (p. 11).

For the notion of a *renewed perception of relation*, Gómez-Barris draws on Éduoard Glissant's (1997) conceptualization of relation and difference as a generative force. Glissant promotes a search for opacity (due to complexity) over a transparency (due to simplicity) (Glissant quoted in Diawara, 2017). This prioritizes an aesthetics and composition that always alerts the reader that the process of highlighting one thing necessitates the obscuring of another. The relation amongst things is reasserted even if immediate clarity is sacrificed. Where conventional visualization practice may be geared towards simplification of complexity, Glissant points us in another direction—to maintain the complexity and not succumb to the lure of absolute meaning. Glissant urges us to move towards poetry because of the generative and irreducible character of its construction and performance.

Complexity and the production of difference is a result of myriad connections and relations of varying strength, duration and character. The emphasis on relations is the counter to extractive logics. This direction may produce a messy problem for dominant notions of visualization, which have been obsessed with visibility as a form of bringing light to a poorly understood reality. In this way, the dominant notions reproduce an Enlightenment agenda that builds a correspondence between unity, creation of the map, employed a protocol by which she made contact with all constituents who were custodians of traditional knowledge of a given place. This meant that trust between visualization designer and participant-as-source-of-data needed to be established and cared for. The orality-based community knowledge of place was in part represented through traditional names—names that rarely, if ever, appear in maps but instead in stories retold over millennia.

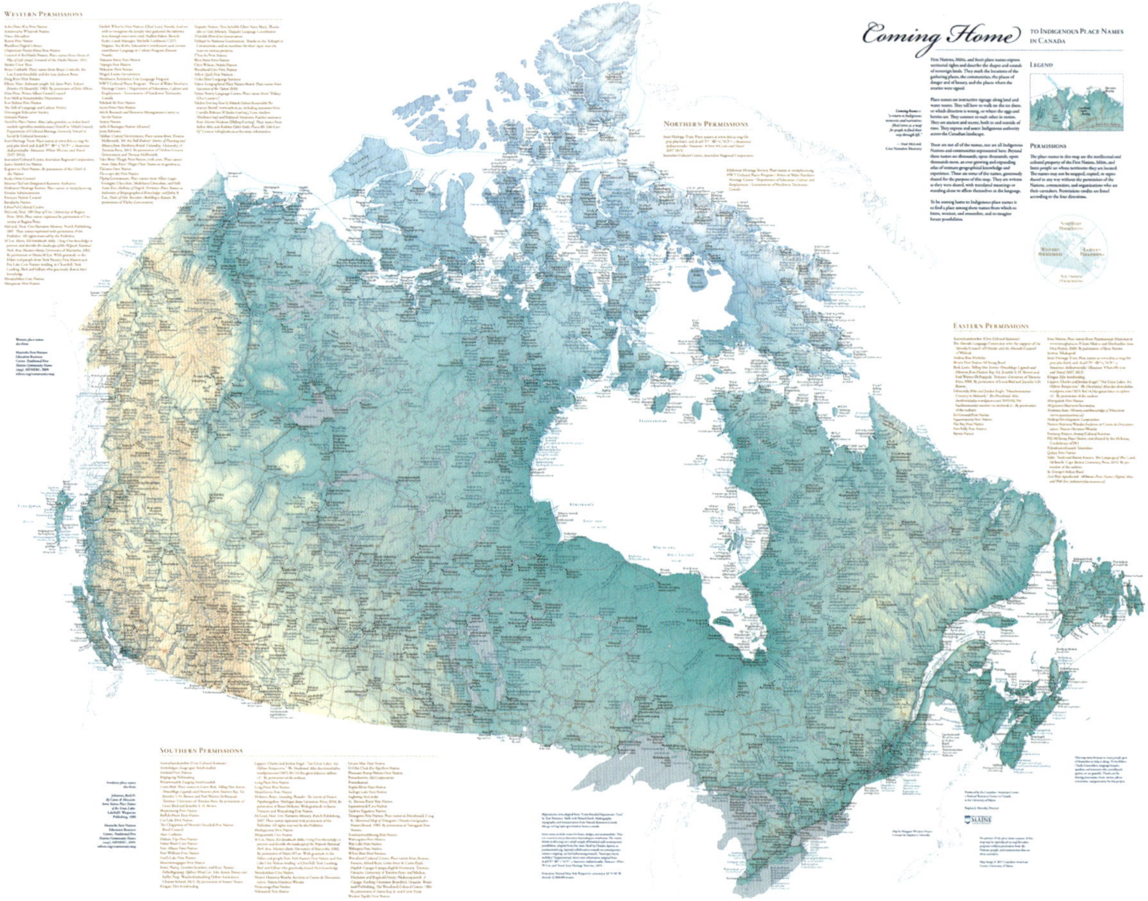

Figure 7.01—***Coming Home to Indigenous Place Names in Canada* (2017)**. Map design © 2017 Canadian-American Center, University of Maine. Place names remain the property of the First Nations and Indigenous communities whose names they are.

Pearce hence does the work of *vincularidad*. Mignolo and Walsh write, 'incularidad is the awareness of the integral relation and interdependence amongst all living organisms (in which humans are only a part) with territory or land and the cosmos. It is a relation and interdependence in search of balance and harmony of life in the planet' (2018, p. 1). Our interdependence requires accountability in our design processes. Pearce's method embeds this logic and further directs the project to keep doing the work of telling the story of interdependence between humans, non-humans and land.

Beyond Critical Visualization Practice

Many critical visualization projects embed an educational purpose. As we saw, Beehive Collective expressly uses its large and incredibly detailed maps in storytelling and teaching. Iconoclasistas finds collaborative methods of mapping that both rely on existing participant knowledge and sharing of knowledge. Rather than treat the participants as empty vessels where you can deposit information (i.e. the 'banking' concept of pedagogy), these projects emphasize reciprocity and acknowledgement of the lived experience and position of the participant. Freire's (1968/2017) highly influential notion of critical pedagogy, opposed the 'banking' model of teaching/learning and instead promoted the co-production of knowledge and acknowledging the authority of the teacher as a tension in the classroom.

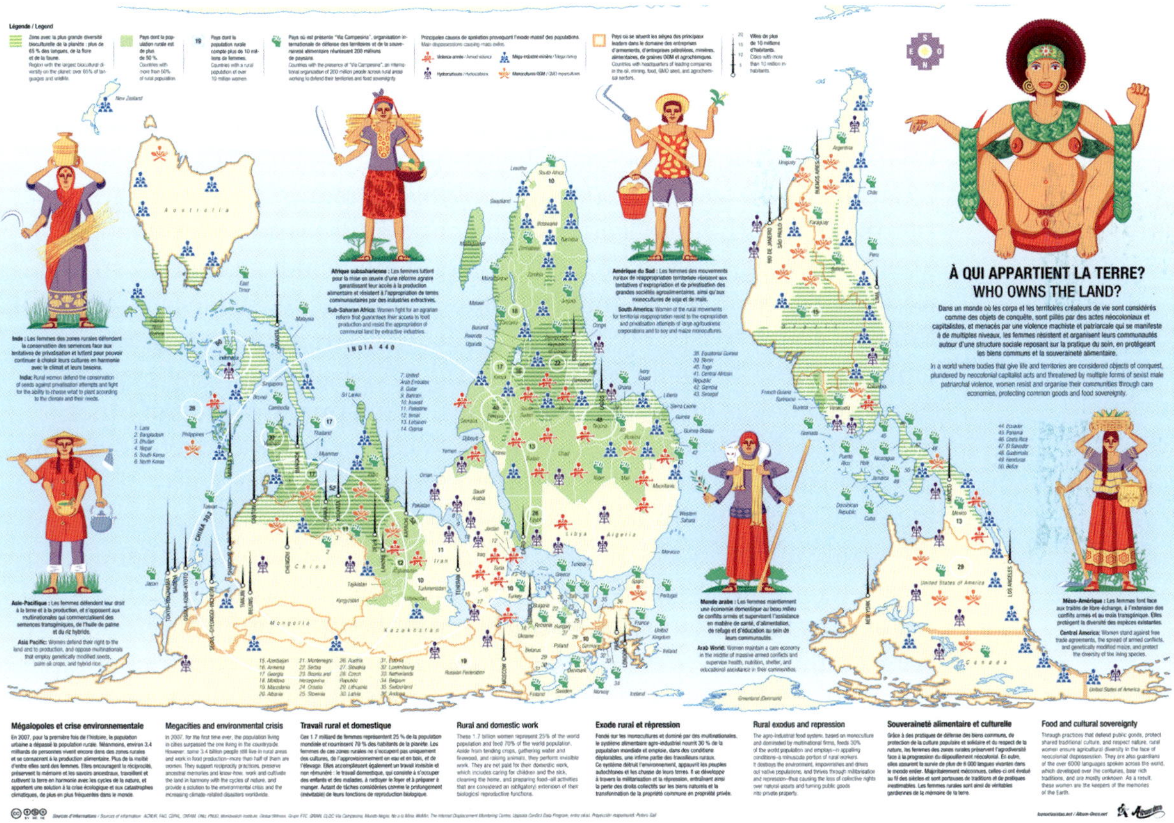

Figure 7.02—**Who owns the land? (2017).** Iconoclasistas

Janet Csontos' project—Land Acknowledgement: Place Based Histories and Indigenous Leadership—works to build knowledge collectively through the addition of image, text and story to specific locations on Turtle Island (2019). Csontos uses the visualization as a tool to let a knowledge sharing and validation process emerge. The map served as a proxy for land and aided her student-centred approach, which directly linked history to territory rather than linear time. She states that, 'Inspired by Vine Deloria's writings, the shift from time-based history to place-based history was intended to evoke a more concrete understanding (than what would be learned from memorizing events on a timeline) of Indigenous peoples' profound connection to the lands here, and of Canada's colonial history' (Csontos, 2019). Csontos' visualization project mediates a process within a larger process of decolonization. Within a settler-colonial context like North America, this requires a restitution of territory to the Indigenous peoples who continue to live here. It is important to point out this specificity due to the increasing use of the word 'decolonization'. As Eve Tuck explains, 'Land is what is most valuable, contested, required. This is both because the settlers make Indigenous land their new home and source of capital, and also because the disruption of Indigenous relationships to land represents a profound epistemic, ontological, cosmological violence' (Tuck and Yang, 2012, p. 5). Land and territory is a source of traditional and spiritual knowledge and thus is part of how a people survives.

When advocating for a decolonization, as non-Indigenous authors, we are also careful to not reproduce the settler logic described by Tuck and Yang. 'The settler intellectual who hybridizes decolonial thought with Western critical traditions (metaphorizing decolonization), emerges superior to both Native intellectuals and continental theorists simultaneously' (Tuck and Yang, 2012, p. 16). This is a profound challenge for us, in that we solidarize with decolonial Indigenous movements and learn from their thinkers and activists, but need to reflect very carefully on how we participate in these struggles from the outside. As Tuck and Yang explain, decolonization is not a metaphor, 'it is not an approximation of other experiences of oppression' (2012, p. 3). When we think of decolonization in visualization practice, we are attempting to articulate a way of supporting these struggles through terms established by those most affected and those who are members of these communities. We recognize that decolonization may mean something different in other contexts of territory and history. As visualization practitioners, it is therefore our responsibility to these communities to reflect their priorities in their particular conditions. When representations of territory have been instrumentalized to circulate forms of knowledge of a space, it has been historically in the service of colonial governments and corporate interests where useful information is encapsulated in calculable form and easily communicated to decision-makers far away. Visualization and mapping did not originate in modernity, but our experience of these techniques is very much rooted in modernity. Modernity is characterized by a commitment to rationality, transparency, and universalism—visualizations have sought to aid in this pursuit. Maps translate an unfriendly, dense, lively and overwhelming territory into a bird's eye view that is readily graspable due to its stable form.

Haraway calls these views provided by visualizations *apollonian* after the Ancient Greek God Apollo, associated with reason, culture, harmony and restraint as well as with the sun that travels across the sky. It is this distance from above and the emphasis of reason that Haraway ascribes to dominant forms of described reality. Haraway (1988) calls this the god-trick, the infinite view from nowhere, innocent, not attached and therefore guaranteeing critical distance and objectivity.

As Cosgrove (2003) reminds us, the Apollonian eye, the disinterested view from above earth also promotes a view of the world as a unified whole. A pluriversal approach works against this one world narrative that invites hierarchical order and mastery.

As an antidote to the totalizing vision, Haraway offers a partial vision, situated in the observed, aware of its generativity and therefore responsible and accountable to what is seen. This is, in part, what informs the multiple worlds as non-exclusive forms of understanding reality. While the multiplicity of positions and knowledges may conjure images of relativism, it is ultimately a false association. Haraway points to how relativism guarantees the ability to be everywhere by reiterating a connection to nowhere. Instead, she states, 'The alternative to relativism is partial, locatable, critical knowledges sustaining the possibility of webs of connections called solidarity in politics and shared conversations in epistemology' (Haraway, 1988, p. 191). Thus, a critical visualization practice can operate by prioritizing subaltern, marginalized, counter-hegemonic, situated knowledges that acknowledge their partial views, by acting as an ally to those who maintain these epistemologies.

In the pursuit of revealing the wages of modernity, or colonialism or capitalism, it has often been the case that the role of visualization has been devoted to the presentation of evidence of this violence and history of extraction. While these efforts are necessary, they do not represent the only task required of visualization. The question of what authority is being appealed to with the representation of evidence, and whether that status quo will change as a result, is especially pointed within colonial conditions where the authority has a vested interest in maintaining the current relations of power. Other directions for representation with data (and other material) can be imagined that work outside of this limited economy. Tuck (Tuck, 2009; Tuck and Yang, 2012) provides a useful framework to think beyond too-narrow a logic of criticality by articulating the difference between pain-centred research and desire-centred research.

Tuck identifies two forms of committed scholarship that have two very different outcomes. Damage-centred research, Tuck says, is 'research that operates, even benevolently, from a theory of change that establishes harm or injury in order to achieve reparation' (Tuck, 2009, p. 413). This approach uses a theory of change, of reparation, that stems from a litigation framework where providing evidence of wrong-doing will bring about justice (Tuck, 2009, p. 413). Visualization practices, as forms of witnessing and providing evidence, often work in this mode. Through the presentation of new data in forms that reveal new patterns, it is hoped that either a critical mass of viewers, or specific authorities, will finally see the truth and therefore act in the service of justice.

The second form of committed scholarship identified by Tuck (2009) is 'desire-centred research'. Tuck describes desire as 'a thriving of the dichotomized categories of reproduction and resistance' (Tuck, 2009, p. 419), where reproduction is the affirmative and the reiteration of the status quo, and resistance is the opposition to oppressive forces. The question that desire posits is how does the work help heal and develop a futurity of a community? Tuck states that, 'Desire, yes, accounts for the loss and despair, but also the hope, the visions, the wisdom of lived lives and communities. Desire is involved with the not yet and, at times, the not anymore' (2009, p. 417). The work of Pearce and Csontos are natural examples of visualization that operates through desire, healing and futurity.

This also resonates with what a graduate student once told us about her own mapping thesis project concerning the colonial process in Canada. As a settler, she had come to understand her implicatedness in the dispossession of First Nations, specifically in the territory in which she had grown up. As with many territories in Canada, Indigenous communities were progressively removed from their land, either through interpretations of treaties, structural or military violence, or contracts made in bad faith. To present and receive feedback on her archival research detailing a specific case of dispossession, she held visiting hours at a community centre. Residents would wander in to peruse the materials and maps and sometimes enter into a discussion. On one occasion, she talked with a community elder about the role of maps. After looking through the visualizations and tracing the history it revealed, he suggested that these maps could be a tool for healing. Healing through the presentation of the roots of intergenerational pain, as well as through the sharing of experiences.

The function of critical visualization as a decolonizing tool/process needs to reflect its context. Decoloniality is not one universal term from the outside. On Turtle Island and in a settler-colonial context, it is directly linked to sovereignty and restitution of territory. In other, interrelated contexts, such as Latin America, this may also include both Indigenous and mestizo epistemologies marked by modernity. Decolonial can be understood not as post-colonial—focused on the various histories of colonies that gained independence—but rather an interrogation and disruption of the continued colonialist relations between powerful states and Indigenous peoples, and the cultural logics that reinforce these relations. Mignolo and Walsh remind us that 'coloniality is constitutive, not derivative, of modernity' (2018, p. 4). Colonization has also meant a continued asymmetrical relationship between the Global South and the Global North. Joaquín Torres-García's *América Invertida* signals this by imagining our sense of North-South as corresponding to up-down and correlating to good-bad. Alfredo Jaar's *A Logo for America* similarly disrupts US-centric naming conventions and reminds us of the presence of South and Central America in the designation 'America'. Reminiscent of associations of white and black to good and bad, and reproduced and applied by white supremacist thinking to skin colour, the up-down/good-bad/master-slave dichotomies play out in our geographical imagination. Iconoclasistas employs this now-famous gesture as a pedagogical move in their mappamundis showing Indigenous and feminist resistance to environmental violence.

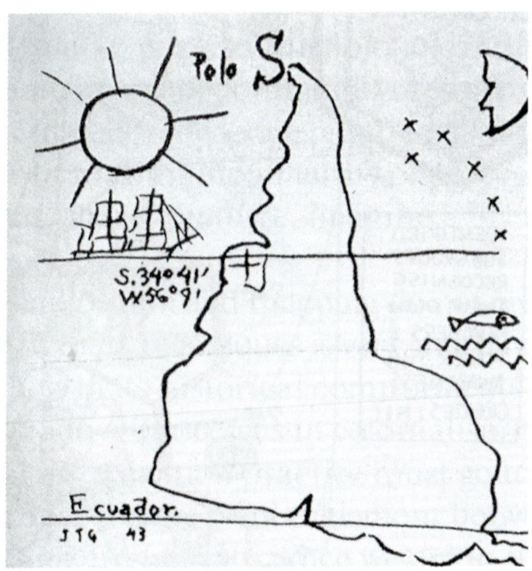

Figure 7.03—***América Invertida***. Joaquín Torres-García

Lauriault, T. P., and Francoli, M. (2017). Openness, Transparency, Participation. In R. Kitchin, T. P. Lauriault, and M. Wilson (Eds.), *Understanding Spatial Media* (1st edition). SAGE Inc.

Law, J. (2007). *Actor-Network Theory and Material Semiotics*.

Lefebvre, H. (2008). Clearing the Ground. In S. Johnstone (Ed.), *The Everyday*. Whitechapel; MIT Press. (Original work published 1961).

Lefebvre, H., and Lefebvre, H. (1991). *Critique of Everyday Life: Foundations for a Sociology of the Everyday*. Verso. (Original work published 1977).

Leonelli, S. (2014). What Difference does Quantity Make? On the Epistemology of Big Data in Biology. *Big Data & Society*, 1(1), 205395171453439. https://doi.org/10.1177/2053951714534395

Levering Lewis, D., and Willis, D. (2010). *A Small Nation of People: W.E.B. Du Bois and African American Portraits of Progress*. Zondervan.

Liu, Z., Nersessian, N., and Stasko, J. (2008). Distributed Cognition as a Theoretical Framework for Information Visualization. *ieee Transactions on Visualization and Computer Graphics*, 14(6), 1173–1180. https://faculty.cc.gatech.edu/~john.stasko/papers/infovis08-dcog.pdf

Lobo, R. (2014, January 21). Could Songdo be the World's Smartest City? *World Finance*. https://www.worldfinance.com/inward-investment/could-songdo-be-the-worlds-smartest-city

Lombardi, M. (2004). *Mark Lombardi: Global Networks* (R. C. Hobbs, Ed.). Independent Curators International.

Loos, A. (2002). Ornament and Crime. In B. Miller and M. Ward (Eds.), *Crime and Ornament: The Arts and Popular Culture in the Shadow of Adolf Loos* (pp. 29–36). YYZ Books. (Original work published 1908).

Lorde, A. (2018). *The Master's Tools Will Never Dismantle the Master's House*. Penguin Books. (Original work published 1984).

Loukissas, Y. A. (2016). Taking Big Data Apart: Local Readings of Composite Media Collections. *Information, Communication & Society*, 20(5), 651–664. https://doi.org/10.1080/1369118X.2016.1211722

Lozano-Hemmer, R., Barrios, J. L., and Labastida, A. (2015). Conversation on the Layout of the Exhibition. In *Rafael Lozano-Hemmer: Pseudomatismos* (pp. 184–221). MUAC, Museo Universitario Arte Contemporáneo, UNAM. https://www.lozano-hemmer.com/texts/bibliography/monographs/catalogue_pseudomatisms.pdf

Lupton, D. (2016). *The Quantified Self: A Sociology of Self-tracking*. Polity.

Lynch, M. (1991). Science in the Age of Mechanical Reproduction: Moral and Epistemic Relations between Diagrams and Photographs. *Biology & Philosophy*, 6(2), 205–226. https://doi.org/10.1007/BF02426838

Lyotard, J.-F. (2005). *The Postmodern Condition: A Report on Knowledge* (Repr). Manchester University Press.

MacEachren, A. M. (2004). *How Maps Work: Representation, Visualization, and Design* (Pbk. ed). Guilford Press.

Magnello, M. E. (2012). Victorian Statistical Graphics and the Iconography of Florence Nightingale's Polar Area Graph. *bshm Bulletin: Journal of the British Society for the History of Mathematics*, 27(1), 13–37. https://doi.org/10.1080/17498430.2012.618102

Magnet, S. (2011). *When Biometrics Fail: Gender, Race, and the Technology of Identity*. Duke University Press.

Mann, M. E. (2014). *The Hockey Stick and the Climate Wars: Dispatches from the Front Lines* (Paperback edition). Columbia University Press.

Mann, M. E., Bradley, R. S., and Hughes, M. K. (1999). Northern Hemisphere Temperatures during the Past Millennium: Inferences, Uncertainties, and Limitations. *Geophysical Research Letters*, 26(6), 759–762. https://doi.org/10.1029/1999GL900070

Manovich, L. (2002). *The Language of New Media*. MIT Press.

Marcos. (2004). *Ya Basta! Ten years of the Zapatista Uprising* (Ž. Vodovnik, Ed.; 1st ed). AK Press.

Markley, R., and Cohen, T. (2012). Time. In *Telemorphosis: Theory in the Era of Climate Change*. Open Humanities Press.

McCandless, D. (2012). *Information is Beautiful*. Collins.

McCandless, D. (2015). *Welcome to our Office: David McCandless*, renowned data journalist and speaker—YouTube. https://www.youtube.com/watch?v=BkQEUJn86MU

McDonald, L. (2006). *Florence Nightingale on Society and Politics, Philosophy, Science, Education and Literature: Collected Works of Florence Nightingale, Volume 5*. Wilfrid Laurier University Press.

Menking, A., and Rosenberg, J. (2020). WP:NOT, WP:NPOV, and Other Stories Wikipedia Tells Us: A Feminist Critique of Wikipedia's Epistemology. *Science, Technology, & Human Values*, 016224392092478. https://doi.org/10.1177/0162243920924783

Mignolo, W., and Walsh, C. (2018). *On Decoloniality: Concepts, Analytics, Praxis*. Duke University Press.

Milner, Y. (2018, April 7). *An Open Letter to Facebook from the Data for Black Lives Movement*. Medium. https://medium.com/@YESHICAN/an-open-letter-to-facebook-from-the-data-for-black-lives-movement-81e693c6b46c

Minard, C.-J. (1862). *Des tableaux graphiques et des cartes figuratives*. impr. de Thunot.

Miranda, C. (2017, January 27). *Artist Theaster Gates on W.E.B. DuBois and what Donald Trump doesn't get about Chicago—Los Angeles Times*. https://www.latimes.com/entertainment/arts/miranda/la-et-cam-theaster-gates-regen-projects-20170127-story.html

Moere, A. V., and Hill, D. (2012). Designing for the Situated and Public Visualization of Urban Data. *Journal of Urban Technology*, 19(2), 25–46. https://doi.org/10.1080/10630732.2012.698065

Mollerup, P. (2015). *Data Design: Visualising Quantities, Locations, Connections*. Bloomsbury Academic, an imprint of Bloomsbury Publishing Plc.

Moore, P., and Robinson, A. (2016). The Quantified Self: What counts in the Neoliberal Workplace. *New Media & Society*, 18(11), 2774–2792. https://doi.org/10.1177/1461444815604328

Moro, S. (2021). *Mapping Paradigms in Modern and Contemporary Art: Poetic Cartography*. Routledge.
Morozov, E. (2013). *To Save Everything, Click Here: Technology, Solutionism and the Urge to Fix Problems that Don't Exist*. Allen Lane.
Morton, T. (2013). *Hyperobjects: Philosophy and Ecology after the End of the World*. University of Minnesota Press.
Morton, T. (2018). *Being Ecological*.
Müller, M. (2015). Assemblages and Actor-networks: Rethinking Socio-material Power, Politics and Space: Assemblages and Actor-networks. *Geography Compass*, 9(1), 27–41. https://doi.org/10.1111/gec3.12192
Munster, A. (2009). *Data Undermining: The Work of Networked Art in an Age of Imperceptibility*. New Radio and Performing Arts, Inc. https://doi.org/10.7551/mitpress/8982.003.0007

Neurath, M., and Kinross, R. (2009). *The Transformer: Principles of Making Isotype Charts*. Hyphen Press.
Noble, S. U. (2018). *Algorithms of Oppression: How Search Engines Reinforce Racism*. New York University Press.
Nold, C. (2009). *Emotional Cartography: Technologies of the Self*. Publisher not identified. http://emotionalcartography.net/

O'Neil, C. (2017). *Weapons of Math Destruction: How Big Data Increases Inequality and Threatens Democracy*. Penguin Books.
Oxford English Dictionary. (2020, September). "*aesthesis | esthesis, n.*". https://www.oed.com/view/Entry/3234?redirectedFrom=aesthesis+

Pariser, E. (2012). *The Filter Bubble: How the New Personalized Web is Changing What We Read and How We Think*. Penguin Press.

P

Pearce, M. W. (2014). The Last Piece Is You. *The Cartographic Journal*, 51(2), 107–122. https://doi-org.arts.idm.oclc.org/10.1179/1743277414Y.0000000078
Pearce, M. W. (2017). *Coming Home to Indigenous Place Names in Canada* [Map]. Canadian-American Center, University of Maine. https://umaine.edu/canam/publications/coming-home-map/coming-home-indigenous-place-names-canada-pdf-download/
Perec, G. (1997). *Species of Spaces and Other Pieces* (J. Sturrock, Ed.). Penguin Books. (Original work published 1974).
Perry, C. (2010). Review of Heath Bunting 0–9 The System Deluxe, Helix Arts, 2009. *Art Monthly*, *334*, 30.
Pevsner, N. (2005). *Pioneers of Modern Design: From William Morris to Walter Gropius* (4., rev.expanded ed). Yale University Press. (Original work published 1936).
Pickles, J. (2004). *A History of Spaces: Cartographic Reason, Mapping, and the Geo-coded World*. Routledge.
Pickles, J. (2006). On the Social Lives of Maps and the Politics of Diagrams: A Story of Power, Seduction, and Disappearance. *Area*, *38*(3), 347–350.
Plasa, C. (2014). Ekphrastic Poetry and the Middle Passage: Recent Encounters in the Black Atlantic. *Connotations*, *24*(2), 290–324.

Playfair, W. (1801). *The Statistical Breviary: Shewing, on a Principle Entirely New, the Resources of Every State and Kingdom in Europe; Illustrated with Stained Copper-plate Charts the Physical Powers of Each Distinct Nation with Ease and Perspicuity: To which is Added, a Similar Exhibition of the Ruling Powers of Hindoostan.* T. Bensley, Bolt Court, Fleet Street.

Polak, E. and Waag Society. (2002). *Realtime Amsterdam.*

Porter, T. M. (1995). *Trust in Numbers: The Pursuit of Objectivity in Science and Public Life.* Princeton University Press.

Powell, J. (2019). Google Releases New Tool to Gauge Pollution. *China Daily Global.* https://global.chinadaily.com.cn/a/201910/11/WS5d9fd824a310cf3e3556fd94.html

Price, B. (1992). Frank and Lillian Gilbreth and the Motion Study Controversy, 1907–1930. In D. Nelson (Ed.), *A Mental Revolution: Scientific Management since Taylor* (pp. 55–78). Ohio State University Press.

Quantified Self. (2018, September 22). *Be An Exhibitor.* QS18 Quantified Self Conference. http://qs18.quantifiedself.com/exhibitor/

Rabinbach, A. (1992). *The Human Motor: Energy, Fatigue, and the Origins of Modernity.* University of California Press.

Rancière, J. (2010). *Dissensus: On Politics and Aesthetics* (S. Corcoran, Trans.). Continuum.

Ratti, C., and Claudel, M. (2018). SENSEable City—Conversation. In M. Del Signore and G. Riether (Eds.), *Urban Machines: Public Space in a Digital Culture* (pp. 208–214). ListLab.

Rediker, M. (2008). *The Slave Ship: A Human History.* John Murray.

Rekacewicz, P. (2013a, March 1). *The Airport Malls.* Le Monde Diplomatique. https://mondediplo.com/2013/03/12mapmaking

Rekacewicz, P. (2013b, December 22). *Cartes en colère.* Le Monde diplomatique. https://mondediplo.com/2013/03/12mapmaking

Rich, D. (2019, December 1). *Interview with Peter Hall* [Audio call and email].

Ridler, A. (2018). *Myriad (Tulips).* http://annaridler.com/myriad-tulips

Ridler, A. (2019, September 6). *Interview with Peter Hall* [Personal communication].

Ridler, A. (2020, September 23). *Correspondence with Peter Hall* [Personal communication].

Ridler, A., and Pfau, D. (2019). *Bloemenveiling.* Bloemenveiling. https://bloemenveiling.bid/

RNDR. (2019). *Open Highway: A Real Time Light Installation Visualizing Highway Vehicles in Scale 1:1.* https://rndr.studio/projects/open-highway/

Robertson, B. (2020, October 10). *Interview with Peter Hall* [Audio call and email].

Robinson, A. H. (1982). *Early Thematic Mapping in the History of Cartography.* University of Chicago Press.

Rosenberg, D. (2013). Data Before the Fact. In L. Gitelman (Ed.), *'Raw data' is an Oxymoron* (pp. 15–41). The MIT Press.

Rosenberg, D., and Grafton, A. (2010). *Cartographies of Time* (1st ed). Princeton Architectural Press.